# Astronomers' Universe

More information about this series at http://www.springer.com/series/6960

David A.J. Seargent

# Weird Universe

Exploring the Most Bizarre Ideas in Cosmology

David A.J. Seargent
The Entrance, NSW, Australia

ISSN 1614-659X          ISSN 2197-6651 (electronic)
ISBN 978-3-319-10737-0  ISBN 978-3-319-10738-7 (eBook)
DOI 10.1007/978-3-319-10738-7
Springer Cham Heidelberg New York Dordrecht London

Library of Congress Control Number: 2014950888

© Springer International Publishing Switzerland 2015
This work is subject to copyright. All rights are reserved by the Publisher, whether the whole or part of the material is concerned, specifically the rights of translation, reprinting, reuse of illustrations, recitation, broadcasting, reproduction on microfilms or in any other physical way, and transmission or information storage and retrieval, electronic adaptation, computer software, or by similar or dissimilar methodology now known or hereafter developed. Exempted from this legal reservation are brief excerpts in connection with reviews or scholarly analysis or material supplied specifically for the purpose of being entered and executed on a computer system, for exclusive use by the purchaser of the work. Duplication of this publication or parts thereof is permitted only under the provisions of the Copyright Law of the Publisher's location, in its current version, and permission for use must always be obtained from Springer. Permissions for use may be obtained through RightsLink at the Copyright Clearance Center. Violations are liable to prosecution under the respective Copyright Law.
The use of general descriptive names, registered names, trademarks, service marks, etc. in this publication does not imply, even in the absence of a specific statement, that such names are exempt from the relevant protective laws and regulations and therefore free for general use.
While the advice and information in this book are believed to be true and accurate at the date of publication, neither the authors nor the editors nor the publisher can accept any legal responsibility for any errors or omissions that may be made. The publisher makes no warranty, express or implied, with respect to the material contained herein.

Printed on acid-free paper

Springer is part of Springer Science+Business Media (www.springer.com)

*For Andrew, Claudia, Theodore and Persephone (a very new arrival in this weird universe!)*

# Preface

In my first two volumes of the "weird" series—*Weird Astronomy* and *Weird Weather*—I concentrated on what could best be described as anomalous phenomena and observations that did not quite 'fit' straightforward explanations. The next volume in the series—*Weird Worlds*—took a somewhat broader view insofar as recent discoveries about other planets have uncovered many features and phenomena which may certainly be regarded as "weird" or anomalous in comparison with anything experienced on Earth. The present volume continues this approach. Strange observations are certainly included within its pages, but the "weirdness" of this wonderful universe in which we live is definitely not confined to these.

Even to begin explaining the universe as revealed by modern science is to stretch the boundaries of what we normally consider to be common sense. Not only must we somehow get our heads around vast expanses of space and incredible depths of time, but we are also confronted with seemingly contradictory notion such as a vacuum which nevertheless appears to be filled with vast amounts of energy, particles which are also waves, "empty" space which can nevertheless stretch as if it is some kind of expandable fabric and so forth. Most scientists understand these things as being relatively established and non-controversial. But then there are more speculative possibilities such as multiple universes, fundamental entities existing in a hyperspace of many dimensions and the theory that the entire universe is—or can be represented mathematically as being—a giant hologram. Gone are the days when the universe seemed to be a relatively straightforward clockwork system of material particles in motion!

Needless to say, in the face of such a complicated subject as the nature of the universe, not all of the theories put forward over the years have won the general acceptance of the scientific community. Sometimes the reason for this is pretty obvious—the theory simply fails to deliver the goods; fails to give a satisfactory account of observed phenomena. Yet, at other times a theory looks good but simply does not jell with the general line of thinking at the time. Sometimes, theories of this type eventually have their day in the Sun as evidence in their favor mounts and/or attitudes change within the scientific community. A sample of such left-field hypotheses is included here.

Together with the other books of this series, several projects are provided for readers who may like to be more hands-on with this subject. These are, in the main, less astronomical than their counterparts in the earlier works, but that is inevitable considering the nature of the subject matter. Although astronomy is intimately involved with the study of the universe as a whole, the deepest issues cannot be resolved at the telescope alone. The physics laboratory, the computer, and those interesting exercises of reason and imagination that go under the name of thought experiments have equally vital roles to play in unraveling the mysteries of this incredible, beautiful, complex, and wonderfully weird universe!

Let the cosmic adventure begin!

The Universe is not only [weirder] than we imagine, but [weirder] than we *can* imagine.

         J.B.S. Haldane (1892–1964) (altered).

# Acknowledgments

My thanks are extended to Mr. Graham Relf for permission to reproduce his finder chart for Stephan's Quintet (from www.grelf.net) and to the staff of Springer Publishing, especially Ms Maury Solomon, Ms Megan Ernst, and Mr. John Watson, for their help and encouraging support. I would also like to specially thank my wife, Meg, for her continuing encouragement during the writing of this book as well as to all who contributed, in whatever way, to its completion.

No book about this "weird" universe in which we live would be possible without the long and complicated research of generations of scientists, both past and contemporary. To all of these, I extend my gratitude for at least partially unveiling this wonderful realm in which we live, as well as my admiration for their genius and patience in deciphering its message.

Cowra, NSW, Australia                                    David A.J. Seargent

# Contents

Preface .................................................................................. vii

Acknowledgments ............................................................ xi

1. From Water World to Inflation; Humanity Faces
   the Universe ..................................................................... 1
   The First Cosmologists ..................................................... 1
   A Truly Radical Discovery ............................................... 8
      Cosmic Fireworks on a Grand Scale! ......................... 13
      But Lemaitre and Gamow Cannot Be Right! ............. 18
      But Is the Redshift Really About Velocity? ............... 19
      It's About Time (Maybe?) ............................................ 22
      Alice in Redshift Land? ............................................... 24
      Steady States and Continuous Creation .................... 26
      Local Quasars? ............................................................. 38
      Quasars and Hoyle's "Radical Departure"
      from Strict SST ............................................................. 40
      Those Pesky Pigeons! .................................................. 44

2. Space, Time, Relativity ... And Other Things ............. 57
   Richard Feynman's Wagon ............................................ 57
      Scientific Astrology? .................................................... 60
      A Very Lively Vacuum ................................................ 62
   Einstein's Universe ......................................................... 68
      That Old Man Is My Son ... ! ..................................... 72
      ... Or Is He? .................................................................. 75
      The Weird World of Curved Space ............................ 78
      Gravity and Time: Yet Another Time Dilation! ........ 80
      Gravitational Lenses, Black Holes
      and Cosmic Strings ...................................................... 82

xiv  Contents

|  | Not-So-Black Holes?! | 98 |
|---|---|---|
|  | Cosmic Strings | 103 |
|  | Is the Universe a Hologram? | 105 |
| 3. | **The Shape, Size, Age and Origin of the Universe** | **113** |
|  | Has the Universe a Shape and a Size? | 113 |
|  |    Donuts, Axes of Evil and Other Anomalies! | 118 |
|  |    Infinite, Finite … or Both! | 124 |
|  |    Is Infinity a Myth? | 131 |
|  |    A Universe for All Time? | 132 |
|  |    From Whence Came the Universe? | 134 |
|  | The Finely—Tuned Universe; Accident, Design or One-Choice-Amongst-Many? | 141 |
| 4. | **Of Atoms, Quanta, Strings and Branes; Just How Weird Can the Universe Really Be?** | **157** |
|  | Solar-System Atoms and Quantum Leaps: Rutherford & Bohr | 163 |
|  | The Atomic Matrix: Heisenberg & Born | 168 |
|  | What Does It All Mean? | 175 |
|  | Schrodinger's Cat | 177 |
|  | Wigner's Friend | 178 |
|  | How Did Bohr Interpret Quantum Reality? | 180 |
|  | An Intriguing Alternative Interpretation | 189 |
|  | What If the Wave-Function Is Real? | 191 |
|  | "Spooky Action-at-a-Distance" | 194 |
|  | The Universe with Strings Attached | 204 |
|  | The "Big Splat" | 208 |
| 5. | **Observations and Ideas from the Left Field** | **213** |
|  | Odd Redshifts | 215 |
|  |    Strange Companions: NGC 4319 and Markarian 205 | 215 |
|  | Discordant Tunes from the Quintet (and the Sex tet) | 219 |
|  |    The Strange Case of VV 172 | 224 |
|  | Quantized Redshifts; An Even Bigger Headache for Cosmologists! | 227 |
|  | Gravitational Weirdness | 229 |
|  |    Gravity and the Quantum Vacuum | 234 |
|  |    Go to Warp Speed Mr. Spock! | 236 |

But Is the Quantum Vacuum a Real Vacuum
After All!? ............................................................................ 237
The Plasma Universe ........................................................ 242
The Cosmological Model of Thomas Van Flandern .......... 245
The Complex Theory of Relativity:
A Universe of Eons ............................................................ 249
Monster Particles of Little Mass? ...................................... 252
Yet More Gravitational Weirdness? ................................... 253
The "Magic Number"; Deep Mystery
or Just a Coincidence? ...................................................... 256

Appendix: The Tripple Alpha Process;
        An Example of Remarkable Fine-Tuning ............... 261

Author Index ..................................................................... 265

Subject Index .................................................................... 269

# 1. From Water World to Inflation; Humanity Faces the Universe

## The First Cosmologists

From the earliest of times, human beings have asked questions about the world around us. What really *is* this world in which we live? What is its true nature and from whence did it come? Many and varied have been the attempts at answering these questions. Stories of universal genesis, most frequently framed in the symbolism of myth, date to the earliest stirrings of human consciousness, but the first accounts that we might recognize as scientific cosmologies and cosmogonies (i.e. theories of the nature and origin of the universe) come from a school of Greek thinkers living on the Ionian coast of what is nowadays Turkey. The man honored by today's historians of scientific thought as deserving the title of the first known scientist was Thales of Miletus (circa 624–circa 546 BC).

We will come back to old Thales in a moment. First however a few words should be said about the so-called Milesian school in general. The natural philosophers of this school, exemplified by the three outstanding thinkers Thales, Anaximander and Anaximenes, were united in the belief that there was a single substance from which the universe was made. The first and last of the trio identified this basic substance with a familiar form of matter. Anaximander's cosmology was rather more subtle and complex, albeit remaining within the school's overall framework.

Although it is probably pointless to speculate as to why the germ of what we now call scientific cosmology first appeared amongst this group of sages in an ancient Greek colony, the philosopher Alexander ("Sandy") Anderson, son of the prominent Scottish/Australian thinker John Anderson, suggested that their motives may not have been entirely driven by pure intellectual curiosity. Noting that at the time these men were living, Miletus

FIGURE 1.1 Thales of Miletos c624–c546 BC. The first cosmologist (*Credit*: Published by E. Wallis et al. 1875–1879)

was a thriving commercial port where a wide variety of goods were constantly being exchanged, Anderson wondered if, at least at a subconscious level, these philosophers may have entertained the thought that if the basic substance of the universe—the stuff from which all objects were ultimately made—could be discovered, might it not be possible to manipulate this substance in such a way that materials now needing importation from distant lands could be manufactured right at home in Miletus? If everything is ultimately made of (say) water, then the basic substance of wood and gold is the same, i.e. water. Might it not, therefore, be possible to turn wood into gold? Or to somehow process the water of the river straight into gold? If anyone found a way to do that …!

Of course, we have no way of knowing whether such thoughts ever occurred to the Milesian sages, even at a subconscious level, but Anderson's speculation is an interesting one and, ironically, would make these sages spiritual ancestors, not simply of later generations of speculative scientists but equally of alchemists and (dare we say it?) engineers!

Be that as it may, let's return to the founder of the school; Thales.

All we know of his theories comes from a trio of statements preserved by Aristotle and which may be summarized as something like:

Water is the cause of all things,
The Earth floats on water,
All things are full of gods. The magnet is alive because it has the power of moving iron.

The last need not concern us, but the first two statements, when taken together, imply a cosmological system in which water is the ultimate "stuff" of the universe. Although we don't know how Thales arrived at this conclusion, it is not difficult to believe that it was not so much an exercise of pure reason as a deduction made from simple observation. Miletus, we might note, was then situated on the Gulf of Latmos at the mouth of the Meander River. Today, the Gulf is dry land and Miletus is well inland. The water which lapped its periphery in the days of Thales has now turned into solid ground! Maybe Thales heard old stories of areas of the Gulf which, even in his time, had turned into dry land. Indeed, he probably lived long enough to recall from his boyhood how parts of the Gulf had receded, leaving small expanses of solid earth in their wake. And beyond all of this, had he not seen mists rising from the waters (water turning into air) and later falling again as rain (air turning back into water)? What more natural conclusion could there be than that water is the underlying substance of all material existence?

Anderson, however, also raises the suggestion that maybe Thales was really arguing that the chief *property* of water—fluidity—rather than literal water per se, was the real underlying principle of the universe. Everything is therefore "like" or "a form of" water rather than water in the usual sense of the word. A later Greek sage, Heraclitus, stressed the fluidity of the world by comparing it to a constantly changing fire. Although sometimes interpreted as teaching that the world is composed of literal fire, Heraclitus seems rather, to be *comparing* it to a fire in the way that fire is, in one sense, always in motion—flames extinguishing as others burst to life—whilst in another sense remaining constant. Perhaps Thales was saying the same thing. Yet, just as Heraclitus seems vague at times in his distinction between what really is fire and what merely resembles it, perhaps Thales never truly distinguished between literal water and (shall we say) metaphorical water. Looking back over so many years, we simply cannot know. But whatever his exact thoughts may have been, the important issue for us is that in these speculations, Thales launched a new endeavor in human though; scientific cosmology or the employment of observation and reason in the bold attempt to work out the nature of the universe. His conclusions might appear naive and even quaint to us, but the task upon which he embarked remains just as intriguing now as it was in his long-ago day.

Thales' two successors in the Milesian school were Anaximander and Anaximenes. The former was truly an ancient polymath whose range of speculation swept across fields as diverse as astronomy and biology but, in the tradition of Thales, his chief cosmological quest was for the ultimate principle of the universe; the basic substance from which all else is made. Unlike Thales (at least, assuming the traditional interpretation of this thinker's position), he understood this as something other one of the familiar materials of everyday experience. For Anaximander, the ultimate reality of the universe was the "Boundless" or infinite and eternal "primary matter" (if I may use this anachronistic term) from which all forms of observed matter arise and into which they all must eventually disappear.

The third Milesian sage, Anaximenes, in a sense stepped back from the more sophisticated cosmology of his immediate predecessor to something more reminiscent of Thales, i.e. to a position that elevated one of the familiar forms of matter into the role of the basic substance of the world. Yet, in so doing, he also introduced something of vital importance to scientific progress—experiment. Anaximenes reasoned that air could become water and even solid earth through a process of condensation. In short, water is moderately condensed air and rock is air condensed to an even greater degree. On the opposite side of the coin, fire is air that has become more rarefied. That is why the more condensed phases of air settle on the ground while fire always rises upward. Moreover, fire is hot and (Anaximenes argued) this is also easily explained because air becomes warmer as it is increasingly rarefied. Now, according to Plutarch, this is where the experiment enters the scene, and it is one that we can all perform for ourselves even as we read this book (see Project 1).

> Project 1: The Experiment of Anaximenes
>
> To replicate what seems to have been the first recorded scientific experiment, first breathe on the palm of one hand with your mouth open. Does the exhaled air feel warm or cold when it touches your hand? Now purse your lips and blow on your hand. What does the exhaled air temperature feel like this time—warmer or cooler?
>
> Now repeat the experiment in a more modernized form by blowing onto the bulb of a thermometer. What is the result?
>
> On this simple experiment, Anaximenes based his cosmology of compressed and rarefied air.

From the result of this simple exercise, Anaximenes concluded that air grows hotter as it is rarefied and colder as it condenses. (As an aside, Sandy Anderson confessed that he had never heard a satisfactory explanation for the result of this little experiment. The adiabatic process, cooling due to the rapid drop in pressure as the air "compressed" between the lips expands, has been put forward, but Anderson seriously doubted that the small changes in pressure concerned here would be adequate to explain the result. The strength of the flow of exhaled air is clearly not the issue either, as the result is the same even if the air is exhaled strongly through the open mouth and blown very gently between the lips).

The actual speculations of these early thinkers and their successors during the following centuries are of little more than historical interest today. But it is the method of observation, experiment and hypothesis which they employed that stands as their real contribution to our understanding of the world in which we live. The way in which this method is employed may have become more complex and sophisticated over time, but the procedure itself is essentially the same for us as it was for Anaximenes.

In some respects, the Miletian school was a false dawn of scientific cosmology. Certain later philosophers of ancient Greece thought scientific speculation to be a waste of time and, like Alexander Pope many centuries later, decreed the that "the proper study of mankind is man"; especially human morality and political organization. The two greatest Greek philosophers—Socrates and Plato—were especially strong on this point. Their most influential pupil—Aristotle—took a somewhat broader perspective but unfortunately came to be considered by later generations as such an overwhelming authority on just about everything that his own speculations tended to stifle future scientific enquiry rather than stimulate it. Later still, the cosmological model of Ptolemy came to be seen as so self evident as to be open to nothing more radical than a little tweaking here and there; always stopping well short of wholesale revision!

By the way, whatever other mistakes they made, it is not correct to say that Ptolemy and his medieval disciples believed in a small universe. With respect to the fixed stars, he specifically stated that the distance between Earth and these objects is so vast as to make the area of Earth appear as a geometric point by comparison! Because a geometric point is defined as having position but no magnitude, this is tantamount to saying that the distance of the fixed stars is infinite. Ptolemy probably did not mean this quite so literally, but the force of his assertion may still be appreciated. The stars are a very, *very* long way away and the universe is very, *very*, large!

It may be of interest to mention in this connection that the oft-reproduced picture of the intrepid traveler poking his head through the veil of stars to witness the grand workings of the celestial machinery is not, as frequently implied, a mediaeval painting depicting the current and largely Ptolemaic beliefs of the period, but a late nineteenth century woodcut representing the then *popular notion* of what the folk of earlier centuries believed. If belief in the sort of limited universe through which the traveler in this woodcut journeyed ever existed, it was not amongst those who best represented the knowledge of earlier times. Such ideas were no more typical of mediaeval times than those of, say, D. H. Lawrence (whose views about the nature of the Sun were, to put it mildly, unconventional) or William Blake (who believed the Earth to be flat and claimed to have touched the sky with his finger) typify the general opinion of the last couple of hundred years.

Another frequently misunderstood teaching of the cosmology based upon Aristotle and Ptolemy is the supposition that because

FIGURE 1.2 This is *not* what medieval philosophers believed! (*Credit*: Artist unknown. From C. Flammarion's *L'atmosphere: meteorologie populaire*, 1888)

the universe was thought to revolve around Earth, we were somehow the throne of creation. In fact, we were thought more as the sump of creation; the place where gross matter settled, leaving the celestial realms beyond the Moon in a state of unsullied purity. The cosmology of Copernicus and the confirming discoveries of Galileo, in so far as they placed Earth amongst the celestial orbs, actually exalted the position and nature of the Earth just as the discovery of sunspots and other changes in the skies (such as the occasional appearance of nova and supernova and the variability in brightness of certain of the fixed stars) degraded the superlunary realms.

The real revolution in thought that became the birth pangs of the modern scientific era was not the dethroning of Earth and the human beings living upon it, but the realization that we, too, are part of a single universal system, of one material with the stars themselves. This realization was in no small degree fostered by the spread of monotheism, principally in the form of Christianity in Europe but also of Islamic influence in the East. Strictly speaking, the polytheism of ancient Greece did not necessitate a single set of laws governing the entire universe. A multiplicity of gods could imply a multiplicity of "natural" laws! But if the universe is the creation of a single God, a theological foundation is provided for the existence of an all-encompassing set of laws, thereby opening the universe to the scrutiny of a curious human race. In short, if one type of material, governed by a single set of laws, comprises both the earthly and superlunary realms alike, why should the whole thing not be the subject of our study? Paradoxically though, as we shall see as our story continues, this more homely universe has turned out to be weirder than any of the wildest stories of ancient mythology.

## A Truly Radical Discovery

We now skip over the years and centuries to that turbulent year of 1914. As armies prepared for what was to become one of history's most bloody wars, in the far more peaceful environment of a meeting of the *American Astronomical Society* being held in Evenston, Illinois, an astronomer was presenting a paper detailing a discovery that was to prove as radical to our understanding of the universe as any of the many social changes triggered by the War would be to the future of the political world. The astronomer was V. Slipher and his paper detailed his remarkable discovery of a peculiar feature of the spectrum of the so-called spiral nebulae. These objects—sometimes called white nebulae to differentiate them from the gaseous green nebulae that appeared to be essentially great clouds of cosmic gas—had long been known, but their nature had always been a subject of controversy. Once upon a time they were thought to be other solar systems in the making or some other variety of Milky Way denizen, but gradually it became apparent that they were island universes; vast systems of stars at immense distances from our home planet. They were, in point of fact, other Milky Ways—other *galaxies* more or less similar to the one in which we live. Slipher had observed the spectra of a number of these galaxies and what he found was most intriguing. The spectrum of an object enables its composition to be determined by noting the characteristic lines of emission of its constituent atoms, but an object's spectrum can also tell us other things about that object as well. It can tell us whether the object is approaching or receding from us and can even enable its velocity to be calculated. This is possible because the emission lines in the spectrum appear at fixed wavelengths for a source that is stationary relative to the observer. However, if an object emitting the light being examined is moving toward the observer, the light is (in a manner of speaking) "compressed"—its wavelength is shortened—and the emission lines in that object's spectrum appear displaced toward the shorter wavelength or blue end of the spectrum. The greater the velocity of approach, the greater that displacement or blueshift will be. Conversely, for a receding object, the light will be "stretched out" and emission lines displaced toward the red end of the spectrum. Something similar happens with sound waves as well. This is most commonly noticeable in the case of a speeding vehicle with a siren. The tone of the siren perceptibly lowers in pitch as the vehicle passes by. The effect is also apparent when radar is bounced off an approaching or receding object; a technological development beloved by police but hated by speeding motorists!

Astronomers had already noted this *Doppler Effect* (to give it its proper title) in the spectrum of stars. Some of these were found to be approaching, others receding from, Earth. No doubt, when Slipher began his spectroscopic examination of galaxies, he anticipated finding a similar mixture of Doppler red and blue shifts there as well. But he was in for a surprise as, indeed, was the entire astronomical community. For what he discovered was that the overwhelming majority of these objects revealed spectra that was redshifted. In other words, the majority of other galaxies appeared to be rushing away from the Milky Way! The only exceptions were the handful of objects which, together with the Milky Way, comprise the so-called "Local Group", to which nearby galaxies such as the two Magellanic Clouds, the Great Andromeda Nebula and M33 in Triangulum belong.

FIGURE 1.3 A fine example of a spiral galaxy; UGC 12158. The bright star at lower left is a supernova exploding in the outer fringes of UGC 12158 (*Credit*: ESA/Hubble & NASA)

This did not mean that the Milky Way was suffering from cosmic BO or in some way repelling its neighbors. It seemed that most of the galaxies were actually rushing away from *one another*, but the reason for this was obscure.

Slipher, it is recorded, was given a standing ovation at the completion of his paper but unfortunately, his epoch-making discovery seems then to have become largely buried in relatively obscure publications. An extended abstract appeared the following year in *Popular Astronomy* and in the 1917 issue of the *Proceedings of the American Astronomical Society*, but appears to have been overlooked by the wider astronomical community and still less by the scientific community in general. Einstein, clearly, could not have known of Slipher's work when he made his "biggest mistake"

FIGURE 1.4 The Great Andromeda Galaxy M31. This is one of the earliest photographs of this object, taken by Isaac Roberts sometime between 1887 and 1899

FIGURE 1.5 The Triangulum Galaxy M33 (*Credit*: Alexander Meleg)

of introducing a *Cosmological Constant* to cancel one of the consequences of General Relativity—cosmic expansion! Had he been aware of Slipher's finding, he would surely have seen it as a brilliant confirmation of his theory. (Ironically, the Cosmological Constant is back in favor again; not to cancel out cosmic expansion this time, but to explain why it seems to be increasing—but more of that anon).

Be that as it may, the redshift of galaxies remained in a backwater until E. Hubble published his rediscovery of this same phenomenon in 1929. It seems that Hubble was unaware of Slipher's earlier work as he made no mention of it in his own publication (something about which Slipher was not very happy, so the story goes). Nevertheless, as well as confirming Slipher's results, Hubble also found that the redshift of galaxies increased in proportion to their distance from the Milky Way. He derived a coefficient for this cosmic retreat (suitably known as the *Hubble Constant*), the refinement of which has ever since been one of the goals of observational cosmology.

The Slipher-Hubble discovery that the galaxies of the universe are racing away from each other is, quite frankly, weird. But even weirder (at least to our workaday-world-conditioned minds) is the fact that this cosmic expansion is *not*, strictly speaking, a Doppler effect at all! It is certainly *like* a Doppler effect, but there is one important difference. A true Doppler effect, whether manifested in the siren of a speeding police car or in the spectrum of a star within our galaxy, relates to an object moving *through* space at a specific velocity and direction relative to an observer. The cosmic expansion, on the other hand, is a manifestation of the stretching of space itself. It is not that galaxy A and galaxy B are flying away from each other through an existing spatial framework. Rather, there is an increasing distance between them because more intervening space is being created! The conventional model is that of an inflating balloon. As more air is pumped or blown into the balloon, so its skin stretches, in a sense creating more surface area or (in other words) more space. Of course, the model must not be pressed too far. For one thing, no more surface fabric is actually created when a balloon is inflated. It is simply that the already-existing material is stretched out so that it covers a greater area at the expense of decreasing thickness. Again, the expansion only takes place in two dimensions instead of three as in the actual universe. Also, demonstrations of cosmic expansion which add spots painted on a balloon's surface to represent individual galaxies or clusters of galaxies can give an erroneous picture as the spots themselves will increase in size along with the surface of the balloon. This does not happen in the real universe and, as we shall see later, confusion about just this issue has led to at least one mistaken criticism of the entire scenario. Nevertheless, with these warning caveats, the inflating balloon model at least does provide us with some level of visualization of this very counter-intuitive phenomenon.

## Cosmic Fireworks on a Grand Scale!

However accurate or otherwise is the comparison between the universe and an inflating balloon, this cosmic expansion, at least taken at face value, seems suspiciously like the aftermath of a mighty explosion at some time in the remote past. One may say the "explosion to end all explosions" except that it would better be called "the explosion to *begin* all explosions"—along with everything else. In other words, a truly radical (not to say profoundly weird!) model of the universe seemed to emerge from the discovery of universal expansion. Far from being the eternal and globally unchanging cosmos of Victorian science, the real universe gave every sign of being an evolving closed system of finite age; the dispersing debris cloud of a cosmic catastrophe with the individual galaxies as the "shrapnel" of this mighty blast.

That, at least, was the thesis of Jesuit priest and astronomer Monsignor Georges Lemaitre. Lemaitre first aired his theory in a mathematical paper delivered in 1931. Following from the observations of an expanding universe, he drew the logical conclusion that if the universe was now growing larger, the further one looked back into the past, the smaller it must have then been. Based on the data available at the time he presented his paper, it seemed that little more than 1 billion years ago, the entire universe approached the dimensions of a point! Presumably, that was about the time when the great cosmic explosion that marked its birth took place. But what exactly was it that actually blew up at that remote date? What was the nature of the *cosmic egg* that so explosively hatched our universe?

14  Weird Universe

FIGURE 1.6  Georges Lemaitre 1894–1966 (*Credit*: Wikimedia Commons)

Assuming that no matter had come into being during the intervening eons, this cosmic egg must have contained all the matter that exists within the contemporary universe. Everything—from the tiniest speck of dust to the mightiest galaxy—must be made from matter that once belonged to this egg! Lemaitre thought of the cosmic egg as a sort of mega-sized primordial atom. Where and how this atom originated was not his problem. Its existence was taken as a given, although as an ordained Jesuit priest, Lemaitre presumably believed it to have been the creation of God through a process beyond the reach of human scientific inquiry. He did however conjecture that "the notions of space and time would altogether fail to have any meaning at the beginning" and that "the beginning of the [universe] happened a little before the beginning of space and time." This is something always to be borne in mind when speaking of the absolute beginning of the universe, although this present writer would quibble over Lemaitre's statement that the beginning of the universe happened a little before the beginning of space and time. How could anything be before the beginning of time? His intended meaning is clear enough, but it might be better to replace "before" with that more accurate but less-than-lovely phrase "ontologically prior to". In any case, it was at $T=0$ (literally the beginning of time) when the primordial atom exploded in an event that, in the words of Cecilia Payne-Gaposchkin, was a little like what happens "in a very modest (!) scale in an atomic bomb".

Recalling Lemaitre's own words about space and time, it is important to understand that the primordial atom, cosmic egg or whatever else we might like to call it, must not be conceived as existing at some location in an otherwise empty spatial void. Remember that the cosmic expansion is not a Doppler effect caused by objects rushing away from one another through a pre-existing space so much as the creation of more intervening space separating the objects themselves. Taken back to $T=0$, this means that space as well as matter exists in the cosmic egg. Winding time backwards to the beginning, space itself shrinks to a point. Unlike matter though, space is continuously being created, but it is being created as the cosmic egg, in effect, explosively swells as time begins its forward march. To the question "Where was the explosion that birthed the universe?" the answer must be "everywhere". In so far as the contemporary universal expansion is really a continuation of the explosion, it is true to say that we are living *inside* the explosion! Jumping ahead of the story for a while, this is seen by the fact that the radiation from that explosion—itself radically redshifted by the relentless cosmic expansion—has been observed and is found to be coming from every point in the heavens. We can peer out into the cosmic night in any direction that we choose and we will still see the distant glow of the universe's explosive birth. But that, as I say, is jumping ahead of our story!

Back to Lemaitre, his radical model was (like many other radical models before and after) not welcomed too enthusiastically at the time, even though it did offer a straightforward explanation of the observed cosmic expansion. The question of time having a beginning proved to be a sticking point for some and the very notion of a temporally finite universe challenged the widespread belief in the eternity of the material realm.

Then, in 1948 physicist George Gamow, together with his student Ralph Alpher, re-worked Lemaitre's theory of cosmic evolution. Gamow, unlike Lemaitre, conceived the universe as being

FIGURE 1.7 George Gamow 1904–1968 (Credit: Serge Sachinov, photographic reproduction from 3-dimensional work of art)

open and infinite in both time and space, although the universe-as-we-know-it is of relatively recent (insofar as the 1.7 billion years—the best age estimate at the time of his writing—is recent!) genesis. Instead of a primordial atom having finite mass which disrupted at T=0, Gamow proposed that the universe contracted from a state of zero density an infinite time in the past. After contracting for an infinite time, it eventually acquired extreme density but, by the very nature of infinity, remained infinite in extent. (Mathematically speaking, the end state would also have been one of infinite density, but we can be pretty sure that the true value stopped somewhat short of this.) It was at this high density stage that the universe exploded in a cosmic fireball of infinite extent. At first, a hot soup of neutrons, protons and electrons was created; the particles all flying about at such high velocities—a result of the extremely high temperatures then existing—that they could not at that time combine to form atoms. Alpher gave the name of *ylem* (pronounced eye-lem; from the Greek word *hylem* or original matter) to this hot soup of particles. In so far as it was from this undifferentiated state of matter that the universe-as-we-know-it emerged, we can see hear echoes of the Boundless of Anaximander!

After 5 min or thereabouts, temperatures, and therefore velocities, within the ylem had sufficiently subsided to allow particles to collide and adhere to one another. Protons and neutrons collided and combined to form deuterons, i.e. the nuclei of deuterium or heavy hydrogen. For the next 25 min or so, heavier and heavier atoms were built up through this process as the ylem continued to cool and expand, eventually coming to an end through the combined effect of the decrease in neutron numbers (thanks to their decay into protons and electrons) and the increasing separation between the particles themselves. The entire atom-building process therefore took about 25 min; as Gamow expressed it, about as long as it takes "to cook roast duck and potatoes."

Although Gamow himself was apparently skeptical of the enterprise, Alpher and his colleague Robert Herman recognized that the afterglow of this very hot natal state of the universe should be present as an all-pervading warmth throughout space and theoretically should be observable as a microwave background—a sort of dim microwave glow coming from every point of the sky. They calculated that the temperature of space should be 5 K at the present epoch. This prediction, albeit not the actual computed value, was to prove very important some years later, as we shall discuss in due course.

Gamow and his colleagues had hoped that this process could account for the elemental constitution of the entire universe. The hope was that not only the genesis and expansion of the universe could be explained, but that even its elemental constitution could be determined by this single cosmological model. This latter hope proved, however, to be a bridge too far. Although the model performed well in explaining the existence and relative abundance of elements up to Lithium in the Periodic Table, it ran into real difficulties with the heavier ones. It was apparent that something other than the brief process of element cookery following the birth of our phase of the universe was required to explain the synthesis of these.

As we shall see later in this chapter, a satisfactory alternative was indeed found, albeit one which proved to be a double-edged sword for the theory. For the moment however, let us consider a few general points about the Gamow cosmological model.

The model assumes a space that is infinite in extent and either flat (*Euclidean*) or hyperbolic (*Lobachevskian*) and it implicitly assumes a timeline stretching infinitely far into the past and future. The related concept of contraction of the universe from a zero average density at minus infinity to a state at which a catastrophic reversal occurs is difficult to translate from a convenient conceptual model into something that "really" took place. What was this "stuff" that contracted? Elementary particles were not formed until after the high-density, high-temperature phase, so for the infinite time prior to this, the universe must have been composed of a pre-ylem, pre-particulate undefinable WHAT?

Concerning this, Lemaitre appears to have grasped the philosophical issues better than Gamow. The latter could be said to have concentrated more on the creation of the material constitution of the universe, but not so much on the universe itself as a spatio-temporal framework within which that matter exists.

## But Lemaitre and Gamow Cannot Be Right!

At least, that was the cry of many scientists. In the minds of many, the very idea of a universe that began in the relatively recent (!) past seemed anathema. Some even made much of Lemaitre's priestly vocation, claiming that "religious motives" lay behind his model ("scientific creationism" in the true sense of that term!) although that criticism hardly accounted for Gamow, who is known to have been an atheist. Others, while accepting the hot/dense birth of "our" universe, could still not bring themselves to believe that this really was the beginning in any absolute sense and instead proposed a bouncing or oscillating universe of an infinite series of hot/dense explosive births, followed by a slowing expansion and eventual collapse in a fiery implosion which would then serve as the equally fiery birth pangs of the next phase.

The major theoretical problem encountered by the supporters of this latter model involves the loss of momentum at each bounce. Each cycle will become smaller, which in turn implies that the series cannot be infinite. The dreaded initial hot/dense cosmic birth remains, even if it may not have been the hot/dense state that began the cosmic bounce in which we live.

A second major difficulty is observational. Thus, if the observed expansion of the universe is eventually to reverse into a contraction, the recessional velocities of galaxies must be less than the critical value required for eternal expansion. In effect, this is the "escape velocity" of the universe. In a cyclic universe, galaxies must recede slowly enough to eventually succumb to the gravitation break which the mass of the observable universe applies, just as a rocket fired from Earth with less than the escape velocity of our planet eventually falls back to the surface. Comparing the redshifts of distant and nearby galaxies permits astronomers to tell whether the recession is slowing sufficiently for an eventually contraction to occur. This is accomplished by plotting redshifts against the apparent brightness of galaxies of the same type. Assuming that certain galaxies close at hand have similar absolute brightness to their counterparts in the remote universe, astronomers can deduce whether the recession is steady, slowing down or speeding up. If the recession is slowing, distant galaxies will appear brighter (because they are actually closer) than their redshift would indicate *if the rate of cosmic expansion remained constant*. Initial results appeared to indicate that this was indeed the case. The cyclic universe appeared to have the support of observational astronomy, despite the grave theoretical difficulties that it raised. As Nigel Calder expressed it, things were looking bad for Moses but "bully for the Buddha". Nevertheless, as observations of increasingly distant objects accumulated, the picture changed. Moses made a comeback! Indeed, more recent measurements indicate a speeding up of recession rather than a slowing down ... but more about that later.

## But Is the Redshift Really About Velocity?

Other scientists—most notably Fritz Zwicky—tried an even more radical way out of the apparent difficulties raised by the concept of an expanding and evolving universe. They denied that the universe was expanding in the first place, and therefore need not be evolving! The observed redshift in the light of distant galaxies is, they argued, due to a loss of energy over time. As red light is less energetic than blue, light becomes (according to this model) redder as it "tires". This hypothesis has never won very much support, primarily because there has never been any real evidence supporting it. Moreover, it has been argued that, as a "tiring" process must take time, it cannot really be relevant to a photon of light for which (from the point of view of a hypothetical observer riding the photon) time effectively stops! This "time dilation" effect is one of the weird properties of the universe, as we shall later see.

A slightly different attempt to explain the cosmological redshift without the need to resort to an expanding universe was suggested by the controversial writer V. A. Firsoff. As a reviewer of one of his books wrote, this author was "no respecter of orthodoxy"; something of an understatement actually! While we all need our heretics in science, unfortunately some of Firsoff's criticisms of the expansion of the universe were based on misunderstandings—a fact which inevitably weakened his position. In particular, he (mis)understood the expression "expansion of space" to mean that the whole of space—in effect, *every* part of space—shared in the cosmological expansion. In other words, as he understood it, even the space between our ears is expanding! Given this assumption, he is perfectly correct in saying that such expansion could never be measured, for the simple reason that whatever "ruler" or measuring device we might use will be equally affected by the universal expansion. If the space between our ears, for example, really is expanding so that we are all literally becoming swollen-headed over time, it is also true that our hats are swelling at the same rate, so it comes as no surprise that we do not feel them getting tighter! Borrowing and slightly altering William Blake's phrase, Firsoff stated that "a universe in a grain of sand" would appear exactly as our world to its inhabitants. This argument was first used by Bertrand Russell in a totally different context (having no relation to the cosmic redshift) and was simply taken in its entirety by Firsoff and misapplied to empirical cosmology.

However, the expanding universe of Hubble et al. is quite different from Firsoff's caricature. There is no question of "local" space expanding. Space within the Earth, the Solar System, the Galaxy … even within the Local Group cluster of galaxies is *not* expanding. The mutual attraction of gravity dominates the motion of planets within the Solar System, stars within the Milky Way and galaxies within the Local Group. Indeed, the distance between the Milky Way and the Andromeda galaxy—the nearest large system to our own—is shrinking rather than expanding. Andromeda and the Milky Way are actually approaching one another, and will at a very distant date merge into a single giant elliptical galaxy.

Nevertheless, despite his gross misunderstanding of what the term "expanding universe" actually implied in observational cosmology, Firsoff's alternative explanation is nevertheless an interesting one. Briefly, he proposed that photons of light expended energy (became redshifted) through interaction with the gravitational field of the remote masses of the universe. Based on Mach's Principal (about which more will be said in the following chapter of this book), Firsoff proposed that the inertial mass of a particle is due to that particle's being held in place by the gravity of the remote masses. Put simply, the gravity of the distant galaxies tends to cause a particle to remain at rest. To move, a particle must expend energy to overcome this resistance. Even a mass-less photon of light must expend energy to move, and it is this expenditure of energy that we observe as the cosmological redshift.

The universe that Firsoff envisioned was in an even steadier state than that pictured by the Steady State Theory to which we shall soon be turning. His was really a variety of the old idea of eternal matter; a material world which was neither created nor will be destroyed—a non-expanding, non-evolving cosmos in which (through Mach's Principle) the nature of every part is determined by the whole. It is "a self-regulating unit in which all forces and processes exactly balance one another" (*Facing the Universe* p. 183) or what he called a *Le Chatelier Universe* "for by the Le Chatelier Principle the products of any change tend to prevent further change that has produced them".

Firsoff presented his model more as a philosophical speculation than as a detailed and observationally testable cosmological theory. For instance, he made no attempt to work out strict observational tests such as the detailed relationship between redshifts and galactic magnitudes, except for a brief statement that, in a static universe, remote galaxies should appear brighter than predicted in much the way that initial observations indicated (and which, as we have seen, was otherwise interpreted as evidence of a slowing expansion and a cyclic universe).

As we shall see, accumulating observational evidence has increasingly come to support cosmic evolution. Firsoff's Le Chatelier Universe, and the "Machian" redshift that it incorporated, fell at the hurdle of observational astronomy.

## It's About Time (Maybe?)

Whilst on the subject of non-velocity redshifts, mention should be made of an ingenious alternative put forward in the 1930s by British astronomer E. A. Milne (1896–1950). As we shall see in the next major section, it was Milne who first formulated the *Cosmological Principle* that came to play such an important role in later cosmologies.

Initially, Milne considered an expanding universe, but one very different from that of Lemaitre. Instead, he treated the universe as being similar to a group of an enormous number of particles originally distributed at random in a very small spatial region. At first, they are all moving with uniform velocities albeit in different directions. Those which fly away from the group are tugged back into it again by the combined gravitational attraction of the rest. But if the average particle velocity within the collection is high enough, the entire assemblage will expand; the mutual gravitational attraction holding it together diminishes and eventually the group as a whole acts as a system of individual particles expanding outwards with the separation distance between any two particles being proportional to the time of the onset of expansion.

In Milne's early opinion, the galaxies of the universe corresponded to the particles in the hypothesized group. The universe is contained within an expanding sphere, the outermost shell of which is perceived by an observer within the sphere (seeing himself as being at its center) as moving outward at the highest velocity simply because it has been moving for the greatest length of time. The highest velocity at the edge is equal to that of light; the maximum possible velocity according to the Special Theory of Relativity. If the observer at the center of the sphere (which is really not a special place as *each* observer sees himself as being at the center) could actually see out to the edge, he would see the distant galaxies bunched up at the horizon. Special Relativity predicted that time dilates and space contracts as the velocity of light is approached (this will be spoken about at greater length in a later chapter) so for the "central" observer, the distant galaxies will also contract in the direction of their motion, effectively becoming thinner and packing together like piles of deflated vacuum bags, finally merging into a solid wall of "pack-flat" galaxies!

It is interesting to note that in this cosmology, the redshift really is a Doppler Effect. The galaxies are actually moving through (Milne assumed) flat or Euclidean space. Space itself is not being stretched as in the Lemaitre model and its later developments.

Later, Milne put forward another but no less radical cosmology in which the universe was static and the redshift due, neither to velocity *through* space nor to expansion *of* space itself but instead to a peculiarity of time! Maybe, opined Milne, the rate of flow of time has changed over the eons and it is this change that has given rise to the cosmological redshift!

His suggestion depended on there being two kinds of time in the universe. One of these is *clock time* and is measured by macroscopic processes—*clocks* in the broadest sense of the term. Of course, by "clock" he did not simply mean manufactured timepieces. Any regular macroscopic process that beats out time is a clock; the rotation of Earth on its axis, the orbiting Moon no less than a swinging pendulum. The second type of time applies to the atomic world and is marked by the frequencies of spectral lines, radioactive decay of unstable atomic nuclei and the like. Not surprisingly, Milne called this *atomic time*. We do not normally think of these as different types of time, still less that they might not be equal, but Milne raised the interesting speculation that maybe they are not always in tune, so to speak.

The two may arbitrarily be defined as equal at the present moment, but what would happen if clock time is actually slowing down with respect to atomic time? In that situation, a period of time reaching back into the past will contain more years of clock time (*clock years*) than it will years of atomic time or *atomic years*. Taking a period of past time equal to the age of the universe, clock time stretches out to infinity. An infinite number of clock years have passed since the beginning of the universe, even though its age in atomic years is finite.

Put another way, atomic time has been slower than clock time in the past. If this is true, it has some interesting consequences. Because of the finite velocity of light, we know that when we observe a galaxy at a distance of, say, 12 million light years, we are really looking back in time by 12 million years and seeing it as it was then, not as it is today. But if Milne's thesis is correct, this means that atomic time at the epoch that the light left that galaxy was slower than clock time. Now, since atomic time relates to such processes as the vibrations of light-emitting atoms, this implies that the light being emitted by these atoms is at a lower frequency than its present-day counterpart. In other words, it will appear to us as redshifted compared with light of recent origin. Moreover, because the discrepancy between clock and atomic time increases the further we look out into space (and therefore back into time), the redshift increases proportionally with distance.

Both of Milne's cosmological models avoid the difficulty of a very dense state at the beginning of the universe however, like the model of Firsoff, they would be increasingly hard pressed to account for accumulating observational evidence. With respect to the static model of Milne, no real evidence has been forthcoming to justify his views on the nature of time.

## Alice in Redshift Land?

In Lewis Carol's classic, Alice entered Wonderland by shrinking in size. What would happen if we are all like Alice; if we are all shrinking (and no, I'm not talking about weight-loss exercises!)?

Remarkably, just that possibility was proposed in 1931 by Sir James Jeans, one of the fathers of British cosmology. Writing in the science journal *Nature*, Jeans hypothesized that maybe,

> ... the universe retains its size, while we and all material bodies shrink uniformly. The redshift we observe in the spectra of the nebulae is then due to the fact that the atoms which emitted the light millions of years ago were larger than the present-day atoms with which we measure the light – the shift is, of course, proportional to its distance.

FIGURE 1.8 James Hapwood Jeans 1877–1946 (Credit: Uploaded to Wikimedia by Kokorik)

If the atomic mass increases but the charge of the electron remains constant, the latter will jump down to a level closer to the atomic nucleus. This is equivalent to the electrons occupying a higher energy state, so the radiation emitted by these more compact atoms will have a higher frequency (be "bluer") than that from the older and more extended (less energetic) atoms in the distant galaxies. The radiation from these older atoms will therefore be "redder" than that of their modern counterparts. That is to say, the emission lines of atoms of the same elements will today be shifted toward the blue end of the spectrum compared with those of ancient times.

This appears to have been given more in the nature of a speculative possibility than a solid proposal of what is true in nature, although Jeans did not accept the interpretation of the redshift as evidence for an expanding universe. Shrinking atoms provided another alternative, although it is unlikely that anyone holds to it these days, if indeed, anyone every truly did.

## Steady States and Continuous Creation

It is safe to say that none of these alternatives to the Lemaitre-Gamow model stood the test of time, however the late 1940s witnessed the emergence of a model which posed a far more serious challenge and within a decade of its proposal had actually overtaken the earlier one as that favored by the majority of cosmologists. Unlike the tired light and allied models, this powerful theory accepted the expansion of the universe, but the conclusion drawn by both Lemaitre and Gamow—the seemingly "common sense" corollary that the expansion must have had some initial starting point—was rejected. According to this new theory, the universe had always existed in the same form as it does today. Of course, individual stars and galaxies form, evolve and eventually die, but the overall picture remains forever the same. Imagine an eternal forest; not a forest of everlasting trees, but one in which each tree is replaced by a new sapling as it dies and decays, such that any section of the forest always looks the same century after century, millennium after millennium. The trees will be different trees, but the forest remains unchanging. That is the vision of the universe presented by this so-called *Steady State Theory*.

Unlike the hypothetical eternal forest however, there is another factor that the Steady State (let's refer to it as the SST) must take into account. The trees in the forest do not move, but in an expanding universe, galaxies do! If the amount of matter remained constant, there would come a time when all the galaxies except those of the Local Group acquired a velocity of recession, relative to an observer in our own or any given galaxy, greater than that of light. Light from these would never reach this hypothetical observer, who would then find himself in the unenviable position of looking out into an empty universe; empty, that is, except for his own galaxy and the companion members of its home cluster. Worse than this however, given that the SST postulates an eternal universe, this position would have already been reached at an infinite time in the past. And if that is still not bad enough, without further input of matter, the stars in the observer's own galaxies would have used up all their fuel and ceased to exist … an infinite time ago!

There is only one way of avoiding this unwelcome conclusion, but the solution is a truly radical one. Matter is continuously being created! We will look more closely at this suggestion in a little while, but let's spend some more time with this equally radical notion of an eternal, globally unchanging and yet expanding universe.

The theory was first put forward by Cambridge (UK) scientists Hermann Bondi, Thomas Gold and Fred Hoyle about the same time as Gamow and his colleagues were working on their formulation of the cosmic evolutionary theory. Both Bondi and Gold approached the subject from what might be called a philosophical direction (although they may not have been entirely happy with that description!) in so far as they saw the SST as the inevitable conclusion to an extension of the Cosmological Principle. The CP (itself a generalized formulation of the *Copernican Principle* that there is no special place in the universe) was initially formulated in its present form by E. A. Milne in the early 1930s and stated that the universe is both homogeneous and isotropic. In more familiar language, the universe looks the same in every direction. Of course, this assumption (and it *is* a philosophical assumption!) must be taken within the proper context. Obviously, if we from our terrestrial vantage point cast our eyes in the direction of the Sun we have a very different view than if we fix our gaze in the opposite direction out past the outer Solar System. On a somewhat grander scale, our view toward the central regions of the Milky Way galaxy looks very different from that out through the Galaxy's outer rim and beyond into the depths of intergalactic space. Yet, on the widest possible scale—the scale of the universe itself—the view appears equal in every direction, according to the CP.

Powerful though this principle may be, it leaves out one very important factor; the speed of light. Propagating through the vacuum of space at around 186,300 miles (299,800 km) each second, light is extremely fast. Indeed, according to Einstein, it moves at the maximum possible velocity. But it still takes time to travel from A to B, and that means that we never see distant objects as they are *now*. We only see them as they *were* when the light left them. Strictly speaking, we never see *any* object as it is at precisely the present instant. Light takes time—albeit only the most minute fraction of a microsecond—to travel across a room. Even our reflection in a mirror is not completely instantaneous, as light takes time to travel from our face to the mirror and back again to our eyes ... although don't try measuring this with a stopwatch! Nevertheless, when we consider astronomical and (especially) cosmological distances, the non-instantaneous velocity of light does become important. Those of us who are old enough to remember the Apollo Moon landings will probably recall the pauses in conversation between Houston and the astronauts due to the time taken for the radio transmissions to travel from Earth to Moon and back again (radio waves, indeed all electromagnetic radiation, travel at the same speed as light). Although noticeable, this was not a problem due to the relative proximity of the Moon, however it would make having a conversation between Earth and an astronaut on Mars rather tedious! Worse still would be a conversation with someone orbiting the nearest star to our Sun; the faint red dwarf known as Proxima in the Alpha Centauri system. This star is approximately 4.3 light years away, meaning that its light takes that amount of time to reach us. If a question was beamed out to an astronaut in a space ship orbiting Proxima, we would need to wait over eight and one half years to receive the answer! And that is the Solar System's *closest* star!! When it comes to other galaxies, the closest of similar size to our own (the Andromeda "Nebula") is over two *million* light years distant!

It will immediately be appreciated that when we deal with galaxies hundreds of millions of light years away—that is to say, hundreds of millions of years back in time—we are taking in a vista of the universe that is very far from instantaneous. But that is precisely the sort of wide-field point of view required for the CP to be hold. Therefore, to be true to the real universe, the CP must require the universe on the largest of scales to be not just homogeneous and isotropic in terms of (instantaneous) space, but isotropic and homogeneous in terms of time as well. In other words, it must appear the same to any observer not just anywhere but (to borrow a term well used and maybe even coined by philosopher Gilbert Ryle) any*when*. One could also say that this formulation of the CP is implied by relativity theory in so far as this understands time to be a fourth dimension of the space/time continuum.

If the CP holds in three dimensions, there is no obvious reason why it should not hold in the fourth dimension as well.

This extended CP is what Bondi and Gold put forward as the *Perfect Cosmological Principle* (PCP). They saw the SST as being the inevitable consequence of this principle.

The third "father" of the SST—Fred Hoyle—is probably the man most associated with the theory in the popular mind and was arguably its chief and most tenacious advocate over the years. He was almost "Professor Steady State" and let his dislike of the Lemaitre/Gamow theory be known by rather disparagingly remarking that he could not believe that the universe began with a big bang. Ironically, his pejorative expression "big bang" stuck to that theory and was even taken up by its supporters. For ever after, the evolutionary theory of Lemaitre, Gamow and their successors has been known as the "Big Bang Theory" (BBT); a most ironic legacy of Hoyle's.

Although Hoyle ended up at the same destination as Bondi and Gold, he reached it via a different route. Whereas we have dared to characterize the path taken by Bondi and Gold as a philosophical one, Hoyle's was essentially mathematical; a development of the mathematics of General Relativity theory.

To go into the mathematics of Hoyle's SST formulation would be beyond the scope of this book and, let's be honest, the ability of its author. Yet the bare bones of his approach are easy enough to grasp.

Essentially, Hoyle postulated a creation field or *C-field* in addition to the familiar nuclear, electromagnetic and gravitational fields which govern the behavior of matter from the sub-atomic to the galactic scales. The C-field, according to Hoyle, existed on a cosmic scale and governed the creation of matter itself. Matter creation, on this formulation, is a regular and normal aspect of the material universe.

Universal expansion is also a normal aspect of the universe. Hoyle reasoned that as new matter is created, pressure is produced which leads to a steady expansion. In a sense, each newly created particle arrives with a certain amount of new space. On the other side of the same coin, the expansion of space produces a sort of negative gravitational energy. The expansion resulting from this negative energy can be interpreted as work done (i.e. energy expended) in overcoming normal gravitational attraction, and it is this expended energy which is manifested as mass through the famous energy-mass equivalent formula of Einstein; $E=mc^2$ (where E = energy, m = mass and c = the velocity of light). It is this mass, assumed to be in the form of hydrogen atoms, which constitutes the continuously created matter of the SST. The rate at which matter is created, according to this theory, is painfully slow; about one hydrogen atom per century in a volume of space equivalent to that of a large office building. Initially, this continuous creation was assumed to have been uniform throughout the universe. That is to say, the rate hydrogen atoms popped into existence was assumed to be the same here on Earth as in the depths of intergalactic space. This would certainly explain why there is hydrogen between the galaxies (as observed in the absorption spectra of very distant objects), but the amount of hydrogen which uniform creation throughout the whole of space implies would seem to place too much of this material in the intergalactic void. This could be avoided if matter-creation at least partially depended upon the presence of existing matter, such that more new hydrogen would appear in galaxies than in the depths of the void. Introducing variable creation activity had its own complications however ... but more about that later.

Many scientists saw this possibility of bringing the genesis of matter itself within the realm of physical investigation as a great strength of this theory, but advances in astrophysics taking place around the time that the theory was put forward also looked as if the SST might equally explain how and why matter is distributed in the form of various elements. We might recall how Gamow attempted to explain the cookery of the elements during the lifetime of the primordial fireball. We saw how this process appeared relatively satisfactory for the lightest of the elements (up to Lithium) but ran into severe difficulties with anything heavier than this. At first glance, one might be excused for thinking that the SST had an even harder problem accounting for the heavier elements. Although it remained a possibility that matter could be created in some form other than hydrogen, the probability so strongly favored this element that all SST supporters assumed that hydrogen was indeed the genesis element. But if that is true, why is the universe not simply an endless atmosphere of hydrogen?

Supporters of the SST looked to the stars for an answer. Stars, as had long been known, are basically enormous fusion reactors which spend most of their lives radiating energy from the fusion of hydrogen into helium. But why stop there? Is it not equally plausible to think that even the heavier elements can be stellar fusion products, especially considering the violent reactions that take place in catastrophic stellar events such as supernovae? Landmark research by Hoyle, W. A. Fowler, J. W. Greenstein together with husband and wife astronomers G. and M. Burbidge developed a convincing theory for the fusion of heavy elements within stars. This theory, often known as the BBFH (sometimes $B^2FH$) theory from the initials of its chief contributors, has been established beyond any reasonable doubt by subsequent observation. In fact, one very supportive piece of evidence was very swift in coming. In 1952, American astronomer P. W. Merrill identified the element technetium in the spectra of certain stars. The interesting thing about this lies with the unstable nature of this element. Even its longest-lived isotope has a half-life of just 216,000 years; hardly the blinking of an eyelid compared with the age of the universe. This element could not possibly have been cooked in the first 25 min of the Big Bang. It could not have been present in the material from which its host stars were formed but must have been synthesized within these stars themselves through the fusion of lighter and simpler elements, exactly as Hoyle and Co. theorized.

Nevertheless, the BBFH did have an Achilles' heel. Although it worked well for the heavier elements, it was not so good on the lighter ones, especially helium and the hydrogen isotope deuterium. Strictly speaking, if all elements other than hydrogen were manufactured within stars, deuterium should not exist in nature (it is destroyed by the conditions inside stars, not formed there!) and helium should only exist in very modest amounts in interstellar space. However, small amounts of deuterium *are* found in interstellar space and the quantity of helium in space is actually quite large; around 25 % in actual fact—way beyond what the BBFH theory predicted.

These problems were not at first apparent however, and the success, as it was thought in the years immediately following its publication, was seen as providing strong support for the SST.

It was equally true that the BBFH also avoided a nagging problem with the BBT in so far as it avoided the need to cook the entire cosmic elemental menu in the time allowed by Gamow. Nevertheless, it is fair to say that more cosmologists saw it as support for the SST rather than the BBT. That is why, earlier in the present chapter, we said that the way out of Gamow's difficulty regarding the genesis of heavy elements proved to be a double-edged sword for his position. As time went on, one edge of the sword became increasingly blunt as the discrepancy in the light elements helium and deuterium became apparent. Continuing work on the BBT cookery of elements actually predicted an abundance of primordial helium in the order of 25 %, deuterium of 0.01 % and a mere trace of lithium, in pretty good agreement with actual observations. Combining this with the good agreement between the BBFH heavy element synthesis and observation ironically tipped the support in favor of the BBT. But all that came at a later date.

Back in the early 1960s, a consensus of astronomers would most probably have shown more than 50 % favored the SST (the writer is actually aware of one such survey in which the results were something like 60/40 % in favor of the SST. Amusingly, the survey also included a second question as to whether such surveys are of any use. The results were 100/0 % against!). In addition to the original Cambridge trio, very respected astronomers and cosmologists such as R. A. Lyttleton, D. Sciama, the Burbidges. B. Lovell and many more supported the theory.

Nevertheless, there were also some equally respected dissenters in those days, chief of whom was Martin Ryle, head of the Mullard Radio Astronomy Laboratory at Cambridge University. It is significant that the chief opponent should have been a radio astronomer. The two theories could best be tested by observing the most distant reaches of the observable universe. Because of light's finite velocity, the most distant galaxies are observed, not as they are today, but as they were when that light left them and if cosmic evolution is really taking place, by looking far enough into space and backward in time, some evidence of this should be apparent. In short, if the most distant galaxies are different from those nearby, there is reason to believe that the universe is evolving and the BBT or something akin to it is presumably correct. On the other hand, if the distant universe looks the same as the nearby universe (if the PCP holds true) the SST is favored. However, back in the 1950s and early 1960s, optical telescopes did not reach our far enough to give any clear answer one way or the other. Radio telescopes on the other hand, though a far more recent development, had a greater potential for picking up cosmic radio sources at truly immense distances and (consequently) very remote times. It was to these great depths of space—in excess of 3 billion light years—that Ryle's Mullard team reached with their radio telescopes ... and found that the number of radio "noisy" galaxies at these distances was greater than in the nearby universe and significantly in excess of that predicted by the SST. This, according to Ryle and his colleagues, was clear observational evidence that the universe was not the same in the remote past as it is today. In short, it presented clear signs of cosmic evolution, just as Gamow et al. predicted.

Although Ryle's results were challenged at the time, they withstood the onslaught and were eventually accepted even by that staunchest of SST protagonists, Fred Hoyle who wrote in his 1966 publication *Galaxies, Nuclei and Quasars* that "the data show that radio sources were either systematically more frequent, more powerful, or both in the past than they are at present."

Hoyle's reaction to this (for him) unpalatable fact will be looked at in due course, but by then other things were happening as well and the astronomical world stood at the brink of something that future historians of science might compare with the Copernican revolution itself.

Socially, the 1960s is remembered as a decade of change and revolution; not in the violent political sense, but in attitudes and ways of thinking. Astronomically, these years were also ones of revolution, with several discoveries that completely changed our view of the universe and our place within it. The writer recalls an article written, sometime during the middle of the decade, in the journal of an amateur astronomical society in which reference was made to an astronomy textbook written some ten years earlier. The article's author highlighted four or five statements given as accepted and apparently undisputed facts in the book, but which had all been proven wrong since 1960! Tongue in cheek, the author remarked that he would not be surprised if astronomers discovered that the Sun is really a red giant orbiting the Earth!

Although it might not have been quite as extreme as that, a discovery made in 1962 was almost as newsworthy!

By that time, a relatively large number of cosmic radio sources had been catalogued, and, thanks to a program conducted at the Jodrell Bank radio telescope in England, the angular diameter of many of these had been established by the beginning of the 1960s decade. Surprisingly, the diameter of a number of them was found to be very small and it seemed as though these might be some weird type of star (radio stars?) rather than distant extragalactic objects. One of these—known by its catalogue entry as 3C 48—was identified with an optical star (or what *appeared* to be a star), and its spectrum obtained by A. Sandage and T. Matthews. The spectrum was most curious, crossed by lines which appeared to represent unknown elements. The two astronomers also found that the star was slightly variable at optical wavelengths, indicating that it was a relatively small object, small enough for some process to orchestrate variations of light over a short span of time, and therefore probably not very distant.

Although the position of the abovementioned object had been sufficiently well determined, the radio astronomical techniques of the day did not normally permit high precision in this respect. Yet, for objects that look just like faint stars, a precise position must be established if optical astronomers are going to be able to find the right star in a crowded field. Fortunately, some of these radio stars lay in regions of sky crossed by the Moon, and it was this fact that led astronomer Cyril Hazard to try a novel approach to determining their positions. Occultations of stars by the Moon have, historically, been used to fix the latter's exact orbit and the exact timing of the disappearance and reappearance of stars by both amateur and professional astronomers played an important role in the days before radar timing and the like. Through the observation of stellar occultations, the already-determined highly precise position of the star plus the exact timing of the start of finish of the occultation enabled an improved position of the Moon to be determined. Hazard, however, used the better-known position of the Moon's limb, plus an exact timing of the instant that the radio source vanished behind it, to find a suitably accurate position for the radio source itself.

Using this method, Hazard used the Jodrell Bank radio telescope to pinpoint the radio source 3C 212 to within 2 or 3 s of arc and, in 1963, travelled to Australia to employ the Parkes radio telescope in the observation of a predicted occultation of another source, 3C 273. Although the Parkes telescope was the only one in a position to observe this event, conditions were far from ideal. The Moon would be very low in the sky at the time of the occultation; so low in fact that the telescope's dish could not be lowered sufficiently before several tons of metal had been sawn off! For several hours prior to the event, local radio stations sent out appeals that no-one use transmitters during the critical time and all roads in the vicinity of the observatory were patrolled to make sure that no vehicles ventured into the area during the time of the observation. Finally, when the observations were completed and data recorded, duplicate records of the results were carried separately back to Sydney by Hazard and colleague John Bolton, travelling on separate airplanes.

The radio source was seen to coincide with a surprisingly ordinary-looking star in the constellation of Virgo. Long known on photographic charts and catalogues, the optical counterpart of 3C 273 did not look like anything special. There was no sign of the fuzziness associated with galaxies, although deep images did reveal a faint jet-like spike extending away from the "star", superficially a little like the jet extending from the nucleus of the giant elliptical galaxy M87. Coincidentally, this galaxy also lies in Virgo, not very far from 3C 273 in terms of angular distance on the sky. This, however, is pure chance as M87 is a relatively nearby object and 3C 273 is ... well ....

Thanks to the accurate occultation observations, the radio source was found to have two close components; one coinciding exactly with the star and the other with the end of the faint jet emanating from it. That alone suggested that something more than a radio star was involved here, but it was the when the optical spectrum was obtained that things started to look seriously weird. Like that of 3C 48, the spectrum of this object appeared to harbor lines of unknown elements. On closer examination however, the lines were found to be familiar emissions—only shifted toward the red end of the spectrum by a whopping 16 %. If this was a true cosmological shift, this "star" was over two and a

36   Weird Universe

FIGURE 1.9   Quasar 3C 273 and jet (*Credit*: ESA/Hubble)

half *billion* light years away! Now, even more remote galaxies had been observed by the early 1960s, but these all showed up as small and very faint fuzzy spots on long-exposure photographs taken with the 200-in. Palomar Mountain reflector, the largest telescope in the world at that time. By contrast, 3C 273 is bright enough to be seen through the eyepiece of an unsophisticated homemade telescope in an amateur astronomer's back yard. The writer has located it with relative ease in a 10-in. (25.4-cm) telescope using a magnification of 71 and in *Weird Astronomy* suggested that it may even be reached by large high-powered binoculars. Recently, I have attempted this under rural skies using $25 \times 100$ binoculars and on

FIGURE 1.10 Giant elliptical galaxy M87 showing jet emerging from nucleus (*Credit*: NASA)

several occasions noted a tiny pin-point of light marginally visible at the right position. If the redshift of this object truly is cosmological, an amateur astronomer looking at 3C 273 through his or her simple telescope is actually experiencing photons of light that began their journey when the Earth was about half its present age and inhabited only by simple microscopic organisms. Dinosaurs—even trilobites—still lay in the very distant future! To be so bright at that distance, the true magnitude of this object must be almost unbelievable. From a distance of around 30 light years, its brilliance would match that of the Sun—not that we would survive to see it!

Many radio sources turned out to be similar objects to these, and the name *quasar* (for *Quasi*-stell*ar* radio source) was given to them. Their discovery was, by and large, not welcomed by the SST supporters as they all were very distant and therefore hinted at being some type of object characteristic of an earlier evolutionary stage of the universe.

Various attempts were made to get around this problem. Maybe the redshift was not due to velocity, but due to intense gravitational fields? Light does indeed lose energy climbing out of a gravitational well, and this energy loss is manifested as a shift toward the red. Ironically, an argument was raised against this possibility by none other than Fred Hoyle, who drew attention to the fact that certain of the spectral lines found in quasars are those only found in highly rarefied gas. These are the so-called forbidden lines; forbidden in the sense that they cannot occur under natural conditions prevailing on Earth, although they are common in astronomical objects whose light comes from glowing gas at very low density, such as emission nebula and the solar corona. For an object to possess a gravitational field of such strength as to give rise to the types of redshifts observed in quasars any gas would certainly not be rarefied and forbidden lines should literally be forbidden!

Hoyle's position regarding quasars did not lead him to abandon the SST however, although he was forced to modify it in a way which is worthy of noting, even though his "radical departure" (his own words) is now of historical interest only. This we shall return to in a little while, but first we should mention another suggestion put forward by several astronomers in an attempt to get around the troubling matter of the quasar redshifts.

## Local Quasars?

The cosmological problem raised by quasars could conveniently be avoided if their redshifts (or a goodly portion of their redshifts) were due, not to the expansion *of* space but to these object's motion *through* space. In other words, if quasar redshifts were wholly or largely true Doppler effects rather than being due to cosmological recession alone. Some astronomers, most notably, H. Arp, sought to find correlations between galaxies of low or moderate redshift and quasars in an attempt to show that the number of these objects found next to nearby galaxies—as viewed on the sky—is significantly higher than would be expected if no true relationship existed. More will be said in a later chapter about discordant redshifts, but for the present it is enough to say that the claim was strongly disputed from the very start and some of the best examples fell apart on more thorough inspection.

Irrespective of any alleged statistically significant relationship between nearby galaxies and supposedly distant quasars, the suggestion that quasars are speeding away from galaxies runs into a couple of very obvious problems. For one thing, if they are objects (of what type?) that have been shot out of galaxies (how?), surely some would have been propelled in our direction and would be observable as strongly *blue*-shifted quasars! No such creature has been found.

An alternative version might be that all observed quasars were expelled from our own Milky Way galaxy. This may avoid the blueshift, as well as the redshift, problem but what evidence is there that our galaxy experienced (in the relatively recent cosmic past) the sort of monstrous eruption required to throw blobs of matter into intergalactic space with such break-neck velocities? This question might actually be turned on its head by suggesting that the very fact that we are alive and kicking and capable of observing quasars is sufficient proof that an eruption of the required violence did not happen!

Note also that it was said that this alternate version of the galactic-expulsion hypothesis *may* avoid the redshift and blueshift issues. The doubt is raised by the simple fact that if the Milky Way expelled hosts of quasars, we may expect the neighboring Andromeda Galaxy to have acted similarly. Yet, this galaxy is close enough to us for at least some of its quasars to have been observed. As a matter of geometry, quasars between Andromeda and our galaxy would be heading in our direction and therefore showing blue shifts. Once again, evidence for such objects is completely lacking. Nevertheless, it would be true to say that until quasars were definitely shown to be at the distances indicated by their redshifts, the possibility that they might be local in some broad sense remained as a vague specter haunting the whole issue.

We will see below how this specter was finally exorcised to most people's satisfaction, but first let us return to Hoyle and the very interesting direction that his thought followed in the mid-1960s.

## Quasars and Hoyle's "Radical Departure" from Strict SST

Back in the early 1960s—just prior to the first quasar discoveries in fact—Hoyle and his colleague J. V. Narlikar were becoming concerned about the SST's ability to account for such inhomogeneities in the universe as evidenced by large clusters of galaxies. If the creation of matter (as determined by the C-field) is evenly distributed throughout the universe, why do such large "clumps" appear? It seemed to Hoyle and Narlikar that an answer might be found if the C-field was stronger in places of existing matter and weaker in the depths of empty space. This modification did indeed appear to work, but it also suggested that in some parts of the universe, extremely dense conglomerations of matter might form; objects completely unknown, at that time, to observational astronomers. Neither Hoyle nor Narlikar initially concluded that such mega-objects actually did exist in the real universe. Only that they *may* exist somewhere out there. But when the first quasars were found, Hoyle was quick to suggest that these may be the objects that he and his colleague had recently theorized about and, if they were, their actual existence could be a real boost for (at least a modified version of) the SST!

At that time, observational evidence provided by the cosmic redshift of distant galaxies appeared to be tilting in the direction of a closed, cyclical, Big Bang universe. This was to change with future observations, but back in the early 1960s, the observational data base was still quite small and interpretations preliminary. Still, the trend was not what SST supporters had wished to see and (as we have already seen) the concept of a cyclic or "bouncing" universe encountered some serious difficulties of its own. Not the least of these was the very mechanism of the bounce itself. How could a state of cosmic collapse be turned around into one of cosmic expansion? There seemed no way that known physics could allow such a thing to happen.

It was at this point that Hoyle and Narlikar proposed their most radical innovation. If the strength of the C-field in any part of the universe is determined by the concentration of matter existing there, it follows that the mass existing within a denser region will be expected to increase due to the increase of creation within that region. However, an opposite effect is also working, namely a decrease in local density (and consequently in the rate of matter creation) resulting from the expansion of space. The resulting inhomogeneities can be expected to change over time, but if these changes are slow, a relatively steady-state balance will be preserved. On the other hand, if rapid changes occur (rapid in terms of the cosmic calendar of course!), wide fluctuations from the steady state will result. In the most extreme cases, the creation process within a particular region might even be temporarily cut off. These rarefied regions will expand until matter from the denser surroundings drifts in to fill them. Hoyle opined that the net effect of this would be the establishment within that region of a series of oscillations, mimicking the bouncing universe but without the awkward requirement of a transition through a singularity as required by General Relativity theory. The emerging theory is of a relatively dense universe existing in an over-all steady state, but with low-density, fluctuating, "bubbles" scattered throughout. Someone likened it to a slab of jelly with air bubbles scattered through it! According to Hoyle, our observable universe exists within one of these bubbles.

Refining his thoughts concerning quasars, Hoyle proposed that these massive objects collapsed during the earlier high-density phase in the life of our bubble. As against the orthodox Theory of General Relativity (i.e. the theory not incorporating Hoyle's C-field), these objects do not collapse into a singularity. Acting like a sort of negative gravity, the C-field increases as the object collapses into higher density states until it eventually overcomes the force of gravity and causes the collapse to reverse and bounce into an expansion. It is this bursting out of material that, Hoyle suggested, is observed as a quasar. As he expressed it "We may well be seeing objects that were typical of an entirely different state of our portion of the universe, of the steady state from which we have departed" (*Galaxies, Nuclei, and Quasars*, p. 131).

Hoyle's radical departure from the strict SST seemed like a way to have the best of both worlds—to account for the observational evidence that increasingly favored the BBT whilst preserving the basic SST. However, the PCP which, we may recall, provided the philosophical foundation for the SST as argued by Bondi and Gold, had to be surrendered unless one insisted that it applied to a region of the universe significantly larger than our own bubble. Copernicus was also shaken, as the assumption that we do not occupy a special place was also compromised. In a real sense, our bubble would constitute a special (though not unique) place. In any event, as observational data on quasars increased, Hoyle's ingenious model failed to stand the test. But Hoyle and the supporters of the BBT were both correct in so far as they saw quasars as representative of an earlier phase of the universe. This became apparent as more and more of these objects were found.

As it turned out, not all quasar-like objects were strong radio transmitters. Many faint "stars" were found to have the extreme redshifted spectra of quasars, yet were silent to radio telescopes. These QSOs (*q*uasi *s*tellar *o*bjects) swelled the ranks even further. It appeared that radio noisy quasars were just one particular type of the broader class of QSO.

Then in 1968, John Schmitt of David Dunlop Observatory found that an object listed for nearly 40 years as a variable star was also a strong radio emitter. This "star" was known simply by its variable-star catalogue title of BL Lacertae. Discovered by Cuno Hoffmeister in 1929, it is characterized by rapid and large variations in brightness, but was not suspected of being anything more exotic than a rather unusual type of variable star within our galaxy. Following Schmitt's discovery however, a spectrum taken of the object showed a large redshift; not quite as large as most quasars, but nevertheless well within that general ball park. Interestingly, the star-like point of light was also found to be embedded in a faint fuzzy spot, about the size and brightness of a galaxy at similar redshift, assuming that the redshift is indeed cosmological. A number of objects similar to BL Lacertae (prosaically-enough known as *BL Lacertae Objects*) are now listed, several of which have their home galaxies visible as fuzzy fringes surrounding the bright stellar point. Although differing from classic quasars and QSOs in their spectra and in the amplitude and rapidity of the fluctuations of their light, they are clearly a type of QSO and their

FIGURE 1.11 BL Lacertae object H0323+022, revealing bright nucleus and surrounding "nebulosity", evidence of host galaxy (Credit: Renato Falomo)

identification as nuclei of very distant galaxies strongly supported a similar identification for other members of the broader class. Much more recently, the host galaxies of some classical quasars have also been identified in very deep images. Indeed, 3C 273 itself has been found to reside at the center of a very remote giant elliptical galaxy.

Earlier, it was mentioned that some doubt as to the cosmic distance of quasars remained as a specter haunting their study. The identification of BL Lacertae objects as very active galactic nuclei of a past cosmic age went part of the way toward getting rid of the annoying ghost, but a far more convincing exorcism was performed in 1970 with the discovery of what came to be known as the *Lyman alpha forest*, i.e. the host of hydrogen absorption lines in the spectra of distant quasars. Briefly, these are lines imprinted on the spectrum of a quasar as its light passes through clouds of neutral hydrogen gas lying between the quasar and the observer at varying distances and, therefore, at different redshifts. The gas clouds themselves belong to the fringes of intervening galaxies. The higher the redshift of the quasars, the more trees are in its forest. The discovery of the forest is just what is expected if the redshifts of quasars are cosmological and was seen by just about everyone as proof of this fact. Quasars are now understood as highly active galactic nuclei, acquiring their power from matter spiraling into a massive black hole at the galaxy's center. For our concern however, the exact nature of these cosmic beasts is less important than the fact that they belong to an earlier age of the universe and their observation through the time machines that we call telescopes yields spectacular proof that the universe is far from being in a steady state. The quasars tell us that we live in an evolving cosmos.

The discovery of quasars would probably have been sufficient to close the door on the SST, however, their message of an evolving universe was not a lone voice. In the year 1965, even as the quasar controversy unfolded, a serendipitous discovery was made that must rank amongst the most important in the entire history of astronomy. Let's now take a look at what this discovery was.

## Those Pesky Pigeons!

About the same time that the first quasars were recognized two scientists, A. A. Penzias and R. W. Wilson, at the Bell Telephone Laboratories, launched an experimental program for finding atmospheric micro-wave noise at 7 cm wavelength. They used an aerial capable of being pointed in different directions and it was found, as expected, that the amount of radio noise detected differed according to the direction of pointing. The degree of noise varied according to the different depths of atmosphere encountered. There were no surprises here, however a small but persistent problem arose in so far as there also appeared to be a faint omnidirectional signal that did not arise from the atmosphere. The signal was clearly real, but it simply should not have been there at all!

The first suspects were pesky pigeons. Having no respect for sensitive scientific instruments, these birds would settle on the aerial and do what birds are wont to do. So the first task was cleaning the aerial of deposits of what Penzias delicately described as "white dielectric material". The birds themselves became unwilling martyrs to science—they were shot! But the mystery signal remained. The pigeons died in vain. They were innocent after all!

After eliminating all other possibilities, just one amazing answer remained. The signal came from outer space. Indeed, not *just* from outer space, but from the most remote reaches of the universe. What Penzias and Wilson had inadvertently done was to take the temperature of the universe!

We recall that Gamow's colleagues, Alpher and Herman, calculated that space should have a temperature of around 5° above absolute zero if the universe had begun in the type of primordial fireball that Gamow's theory implied. Subsequent recalculation had reduced this to 2.7°; the exact temperature required to generate the "nuisance" signal that Penzias and Wilson had recorded. Quite by accident, these two scientists had found the strongest proof to that date that the Big Bang did actually happen! In later years, study of this cosmic microwave background (CMB) would yield invaluable information about the fundamental nature of the universe.

Strictly speaking, the CMB is not a direct observation of the faded remnant of the Big Bang itself, but represents the cosmic era some 372,000 years later when the sea of hot and dense plasma cooled sufficiently for neutral hydrogen atoms to become stable. This is known as the *surface of last scattering* and is the point at which photons could escape the plasma and travel as far as the present-day universe. It is, in short, the era when the infant universe became transparent.

> Project 2: The Big Bang on Television!
>
> Have you ever watched the Big Bang on TV? No, I'm not talking about a certain comedy series! I mean the real thing. Or, at least, the remnant of the radiation from the surface of last scattering.
>
> Alas, with old-fashioned analog TV now largely replaced by digital (in the writer's home country at least) this is no longer possible unless you can get hold of an old set, but can we remember the times when there was a station break or as we changed channels on the old TV sets and the screen filled with thousands of bright sparkles that used to be called "snow"? Remarkable as it seems, between 1 and 2 % of these sparkles—which amounted to quite a large number on a single TV screen—originated at the surface of last scattering of the infant universe. Some of the sparkles you saw when your analog television reception went down were embers of the very fire of creation; surely the most remarkable thing ever seen on television!

The discovery of the CMB and the subsequent observations that have been carried out since 1965 could hardly constitute a stronger argument for the BBT. Nevertheless, the theory, at least in its early forms, was not without its problems.

One such difficulty arose concerning the very uniformity of the CMB itself. Briefly stated, because nothing can exceed the velocity of light and because this velocity is finite, a limit is set to the size of any two regions of the universe which can be in contact. This limit is known as the particle horizon. Now, at the distance of the surface of last scattering, the particle horizons of any point on this surface marks out a circular patch with an apparent diameter of some 2°. In other words, a patch in the contemporary skies some four times the diameter of the full Moon. No physical mechanism could operate over distances greater than 2° at the distance of the CMB. So how can wider regions of the sky at this distance have the same temperature?

A second major puzzle is the so-called flatness of the universe, that is to say, the very close approximation of space on the largest scale to Euclidean geometry. Despite our familiarity with Euclidean geometry, it is actually a highly improbable occurrence. In addition to being flat, space can also theoretically have negative (open or Lobachevskian) and positive (closed or Riemannian) geometries. Indeed, there is any number of degrees of Lobachevskian and Riemannian curvature that space could have. Space may be slightly, moderately or strongly curved in either the positive or the negative sense, each depending upon the density of matter in the universe itself. For high enough densities, space is closed and the universal expansion will eventually slow sufficiently to reverse into contraction and ultimate collapse. If the density is low enough, the universal expansion proceeds sufficiently fast to prevent contraction and collapse. Euclidean geometry, by contrast, can only occur if the average density of the universe is *exactly* enough to prevent contraction. There are no degrees of Euclidean geometry; space is either flat or it is not. And if it is flat, then that is that! No range of flatness makes sense. Strict Euclidean space is a hair-line transition between the closed alternatives and the open ones. In theory, a Euclidean universe will expand to infinity, but the expansion will take an infinite time and when the universe reaches that state, the velocity of expansion will have declined to zero.

Now, we cannot be sure that the real universe is precisely poised on this Euclidean knife edge, but the most accurate observational material shows that space is extremely close to this condition. Even on the largest scales, space appears to be Euclidean for all practical and most theoretical purposes. But herein lies the problem. Even the very slightest initial departure from absolute flatness would be magnified over time. Therefore, if the departure from flatness is minuscule today, it must have been incredibly tiny when the universe was young. In fact, it has been calculated that at the time of nucleosynthesis (when the universe was only a few minutes old) any departure from absolute flatness by as little as one part in $10^{14}$ (let's print this out for maximum impact; one part in 100,000,000,000,000!) would have resulted in a universe either so radically Riemannian or so clearly Lobachevskian as to be totally unlike today's reality. It would either have collapsed long

48  Weird Universe

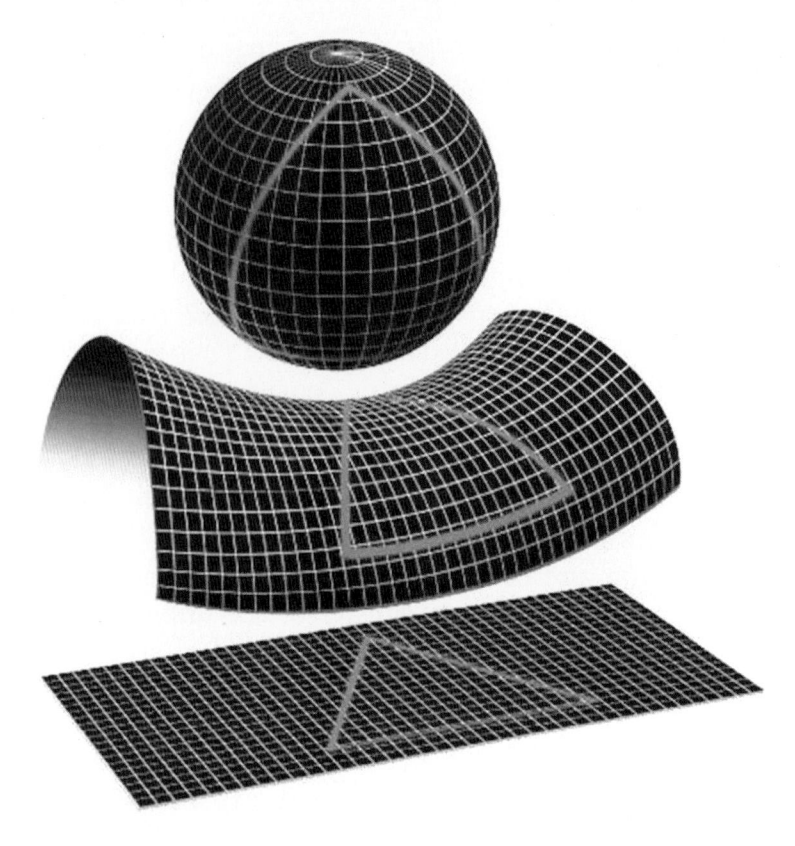

FIGURE 1.12 Three possible geometries of space, Riemannian ("spherical"), Lobachevskian ("hyperbolic) and Euclidean ("flat"). In the first, the angles of a triangle add to >180, in the second to <180 and in the third to 180° (*Credit*: Gary Hinshaw, NASA)

ago (if Riemannian) and we would not even be around to see it, or it would have largely dispersed as rapidly expanding space thrust galaxies along their hyperbolic trajectories into ever increasing obscurity.

Strangely, derivation of the mass of the universe by direct observation of the objects within it appear to favor Lobachevskian space. Because the global geometry of space is governed by the material within the universe, one might think that we should get a fair idea of the geometry of space by estimating the average mass of galaxies and multiplying this by their estimated number. However, taking into account all the visible galactic inhabitants such as stars and nebula, the total mass comes out vastly too small to yield anything other than a strongly negatively curved space. Even taking into consideration dark objects from black holes to planets, the amount of 'ordinary' matter (the type that constitutes familiar astronomical objects) is just 4.9 % of that required for space to be flat. Strangely, most of the universe must consist of something other than familiar matter. That is weird enough, but it does not by itself explain why the universe should be so very nearly balanced on the Euclidean knife-edge.

An Inflating Universe?

Even though most cosmologists, by the latter quarter of the twentieth century, were convinced that the BBT was basically correct, it was clear that our understanding of the very early universe was still far from complete. Some fundamental factor was missing; but what could that factor be?

A very interesting answer was given by American physicist Alan Guth in 1980 and, independently, repeated by Katsuhiko Sato the following year. Both physicists proposed that the universe experienced a (very!) brief period of exponential expansion beginning, according to most estimates, about $10^{-36}$ s after the instant of creation and lasting until sometime between $10^{-33}$ and $10^{-32}$ s from Time Zero (although some recent observations indicate an even earlier time for this epoch; possibly as early as $10^{-43}$ s after Time Zero). Yet, during that time, the volume of space bloated exponentially by the colossal factor of at least $10^{78}$ times! Clearly, space was then expanding at a rate far exceeding that of light, but as no accelerating material mass was involved, Special Relativity was not violated.

Maybe, consciously or subconsciously, Guth associated this burst of runaway expansion of space with the runaway expansion of commodity prices then occurring in the economic systems of much of the world. In any event, he gave it the same title—*inflation*. (Alas, inflation in the economic sense lasted a lot longer than a minute fraction of a second!). Following its burst of cosmic inflation, the expansion of the universe continued at the far more sedate rate that we observe today.

Initially, Guth proposed the occurrence of an inflationary epoch to explain the absence of a weird type of particle that theorists working on grand unification theories of the basic forces of nature had predicted in the late 1970s. These were magnetic monopoles; point-like particles with a single magnetic polarity. That made them very odd indeed, as every experience of magnetism thus far encountered has always involved both poles. Nevertheless, the grand unification theories (rather unpleasantly known as GUTS) then being studied predicted the presence of topological defects in space, arising during the early phases of the universe, which should manifest as these very particles. Yet none had been observed. Perhaps there was some as-yet undiscovered process preventing the creation of monopoles. Maybe these particles were just *too* weird to be real! Alternatively, perhaps monopoles were created but some other process intervened to hide them from our view. The solution proposed by Guth provided a candidate for this second process. Simply put, if the volume of space inflated by a factor of one followed by 78 zeros, the number of monopoles would be so diluted that the likelihood of finding any is effectively reduced to zero.

However, it quickly became obvious that this inflationary phenomenon would, in a similar manner, effectively flatten the universe and so stretch the particle horizon of any point in the early universe that the entire field of the contemporary CMB would indeed have been causally inter-connected. What would have been a single two-degree patch of sky without inflation has now swelled to embrace the entire cosmic background.

The inflationary model also neatly explained the structure that we do observe in the universe—clusters and chains of galaxies and the like. So-called empty space is, according to quantum physics, actually a heaving ocean of fluctuations at extremely tiny scales (more about this later) and, as space exponentially inflated at this extremely early era, the fluctuations then present were likewise stretched into macroscopic—indeed truly cosmic—dimensions. These subsequently acted as the seeds for the growth of the structure that we now observe in the universe. In this way, inflation theory united structure at the smallest sub-sub-atomic scales with that at the largest cosmic dimensions.

For inflation to work, a new field called the *inflaton* field (quantized into inflaton particles) was proposed, but exactly what this might be is not so easily understood. Guth proposed that the universe initially existed in a false-vacuum state. Now, as the word vacuum is used in quantum physics, it is not simply equivalent to nothing. It is the lowest energy state, but because of the prevalence of *Heisenberg's Uncertainty Principle* in quantum analysis (more will be said about this later in this book), this lowest state cannot be exactly specified. But it *would* be exactly specified if it equaled zero! In this rather esoteric-sounding way, empty space is predicted to be filled with fluctuations of energy.

Imagine rolling a table tennis ball down the side of a conical bathroom basin. Left to itself, the ball will end up in the plug hole, which is the state of lowest energy within the basin. However, suppose that we paint a ring of very strong glue around the sides of the basin, say, half way between the plug hole and the basin's rim. If the ball is then rolled down the side of the basin, what will happen? Assuming that the glue is strong enough and the ball light enough, the latter will get stuck half way down the side of the basin. It truly has come to rest, yet has not sunk to the lowest possible energy state. If the plug hole is seen as being analogous to the (true) vacuum, the ball's position held by glue to the side of the basin above the plug hole is an analogue of the false vacuum. Whereas the ball could not be lifted from the plug hole without supplying some energy (lifting it out), a slight picking at the glue will probably be enough to free it and send it rolling the rest of the way into the plug hole. A similar slight alteration of the false vacuum could similarly create a bubble of true vacuum which would then inflate and consume the surrounding false vacuum; and it was this that Guth suggested might have happened during the inflationary era as the infant universe changed from a false to a true vacuum state. (Incidentally, the fear of some scientists that experiments using the Large Hadron Collider might end in disaster had a similar origin. They argued that, although we assume that the universe is now in a true vacuum state, there is no absolute proof of this. If the present vacuum is a false one, the fear was raised that the energies attainable by the LHC might be enough to cause a small bubble of false vacuum to change into true vacuum and undergo its own exponential inflation, swallowing Earth,

Solar System, Milky Way and even the observable universe as it did so! How ever the conditions might change as false vacuum became true, the survival of the world as we know it was far from guaranteed. Other scientists remarked that higher energies occur naturally in the universe and we are still around to witness them. A cosmic ray particle detected in October 1991, for instance, packed an energy 50 *million* times that achievable by the LHC. Moreover, if other human-like species exist out there somewhere, they have probably built their own LHCs, without destroying the universe in the process!)

Increasingly precise mapping of the CMB from suitably placed special observatories, high-altitude balloons as well as from the space-based observatories *Cosmic Background Explorer* (COBE), *Wilkinson Microwave Anisotropy Probe* (WMAP) and *Planck* have greatly increased—and are continuing to increase—our knowledge of the earliest phases of the universe. At the time of writing, Planck data continues to be refined, but the results are broadly in good agreement with inflationary theory. Small ripples within the CMB—the seeds of later development of chains of galaxies—agree with the predicted inflationary stretching of microscopic fluctuations during the first instants in the life of the universe. A major breakthrough came in March 2014 (not this time from Planck, but from analysis of data secured by the South Polar Telescope) with the apparent discovery of another set of ripples in the CMB; these features, if they are confirmed by further observation, are believed to have been caused by the stretching of gravitational waves that could only have occurred during the inflationary era. And in addition to all of this, analysis of the CMB reveals the amazing degree of flatness expected in an inflationary scenario.

Nevertheless, it is still too early to say that inflation solves all the problems. For one thing, there is not a *single* theory of inflation. Inflationary models have appeared on the theoretical cosmology scene like frogs after a storm. Moreover, although the basic predictions of inflation in a general sense appear to describe features of the real universe, other predictions are beyond the capability of observational techniques. Some of the discussion seems more like metaphysics than physics and the line between the two has become difficult to draw. Perhaps it does not even exist.

One of the strong attractions of the model is its ability to deal with the flatness problem and with the uniformity of the CMB. The flatness issue, as we saw, looked very formidable prior to the inflation model. It appeared to demand the universe to be balanced on an extremely fine-tuned line which seemed highly unlikely to have been true in nature. Inflation nicely dealt with this ... or did it! Critics of the model, such as Roger Penrose, argue that it does not solve the fine-tuning issue at all. It only pushes it back a step or two. In other words, for inflation to work in the required way, very strict initial parameters must be assumed. According to Penrose's calculations, the universe without inflation, although highly improbable, nevertheless had a far higher probability of happening than the inflationary universe!

In the first draft of this chapter, the present writer ended this paragraph with the sentence, "Hopefully, further Planck data and continuing analysis thereof will shine more light on this subject". Not long after writing this, analysis of this data has indeed shed more light on the subject, albeit in such a way as to further strengthen Penrose's criticism. The title of a 2013 paper by A. Ijjas et al. sums it up—*Inflationary paradigm in trouble after Planck*. According to Ijjas and colleagues, Planck data reveals none of the inflationary theories to be probable. That does not mean that inflation does not occur, just that it is not something that occurs "normally" but, as Penrose argued, itself requires highly special and fine-tuned conditions. Much more will be said about this in future I predict.

## The Universe After the First 13 Years of Twenty-First Century

As far as we can tell, the universe consists of three components.

First of all, there is ordinary matter. This is the stuff of which you and I, the Earth, the Sun and other stars as well as interstellar gas and dust are made. Although this is the most conspicuous component of the universe (at least for our senses) it only comprises a mere 4.9 % of the universe's mass according to the most recent Planck observations.

The second component is dark matter. This is widely thought to exist in the form of neutrino-like particles which only interact very weakly with visible matter, but which exist in such stupendous numbers as to make their presence very apparent on a cosmic scale. These are generally brought together under the head of *Weakly Interactive Massive Particles* or *WIMPs*. Unfortunately, WIMPs are proving to be very shy. Although several experiments have been set up for the purpose of detecting them, the results have thus far been consistently and annoyingly negative. Despite this, WIMPs remain the most likely candidates although if they continue to elude experimenters, the issue will need to be reconsidered.

Until quite recently, it was thought that dark matter (whether it turns out to be in the form of WIMPS or something else) made up most of the mass of the universe.

But then, another surprise awaited cosmologists. As we have seen, the ultimate fate of the universe—in the sense of whether it will go on expanding or eventually collapse—can be deduced by any change in the rate of cosmic expansion over time. This is not an easy thing to measure, as we can well appreciate. The most reliable way is to find a "standard candle" or some class of object whose members are all of very similar intrinsic brightness as well as being so highly luminous as to be detectable over vast reaches of cosmic space. Such objects do in fact exist. They are a type of very bright supernova known as Type 1a. Supernova are exploding stars and come in several varieties. Some are very large stars that collapse in immense fireballs at the end of their lives, however these are of little use as standard candles as they differ considerably in the maximum brightness that they attain. Type 1a supernova, by contrast, occur when a dense degenerate star known as a white dwarf accretes matter from a distended companion. This process continues until the dwarf reaches a mass of 1.38 times that of the Sun, at which stage a runaway thermonuclear reaction occurs within the accreted blanket of material and the dwarf is blown to smithereens. Because the runaway reaction occurs at a specific mass, the resulting fireball is always of the same brightness. Because of the quantity of material involved, it is also extremely bright. A single supernova of this type can briefly outshine its host galaxy of several hundred million stars.

As the twentieth century drew to a close, a number of astronomical programs conducted searches for these objects and followed up the increasingly frequent discoveries with spectral analysis of the supernova. Assuming that very distant/ancient Type 1a supernova peaked at the same intrinsic brightness as those in the local universe (a very reasonable, though strictly speaking unproven, assumption) it was not difficult to judge their distances and then compare this with the distance that they should have if the cosmic recession of their host galaxies was the same as that of nearby ones. Everyone expected that these results would show that the cosmic recession had slowed down over time—the question on which the fate of the universe depended was by how much? But to the surprise of all concerned, the results consistently showed just the opposite. The rate of cosmic expansion is speeding up!

So here enters the third basic component of the universe. Dark energy—some unobserved energy (maybe the remnants of whatever caused inflation)—is gradually making itself felt as the expanding universe dilutes and the dominance of gravity at the largest scales consequently weakens. As energy also has mass, this also contributes to the total mass of the universe, albeit in a sort of negative way!

The latest from Planck breaks the universe down into 4.9 % ordinary matter, 26.8 % dark matter and 68.3 % dark energy.

By and large, as already noted, the Planck results look good for what has now become the standard model of the universe; Big Bang plus a brief very early burst of inflation. Nevertheless, not everything is plain sailing. The Planck results also contain some anomalies—some departures from what was expected and what was predicted on the basis of the inflationary Big Bang model—hinting that there are still aspects of early cosmic history that remain hidden. Careful analysis of the Planck results reveal an unexpected asymmetry in the microwave sky, weaker than expected fluctuations in the CMB at large angular dimensions (also previously noted by COBE and WMAP) and a mysterious cold spot that was also hinted at by WMAP, but assumed to be artifact which scientists hoped would go away as higher precision results were obtained via Planck. It didn't!

These anomalies may require nothing more than a slight tweaking of accepted theories, but there is also a possibility that they may hint at something more radical. Sometimes what looks like a small anomaly can profoundly alter our picture of reality.

An historical example of just such a "small" anomaly was the precession, or slow secular drift, of the perihelion of the planet Mercury. This was a little larger than Newtonian gravitational theory had predicted. Everything else that the theory predicted withstood the observational tests. The only problem was this slight difference in the orbit of Mercury. It seemed reasonable to astronomers of the nineteenth century to tweak the situation by postulating a small planet orbiting between Mercury and the Sun. Mercury's orbit could then be explained in a straightforward way, according to Newtonian theory, by introducing the weak gravitational tugging by this little planet. So confident were the astronomers of the day in the existence of this object that it was even given a name—Vulcan! Alas for Vulcan and alas for Newtonian theory, this simple tweaking was not the answer. The real solution did not come until Newtonian gravity was replaced by Einstein's theory of General Relativity. The solution was not a new planet, but new physics!

Whether the Planck anomalies are also hinting at new physics or whether there really is a counterpart of Vulcan, in other words, a non-radical solution that leaves the overall picture unchanged, is yet to be determined.

# 2. Space, Time, Relativity … And Other Things

## Richard Feynman's Wagon

Here is a riddle for you. What do throat singers, bongo drums, songs about drinking orange juice, toy wagons and a city named Kyzyl have in com mon?

If you answered "Richard Feynman" you would have been right (Fig. 2.1).

Feynman was one of the previous century's greatest physicists. He was also a hot player of bongo drums, performed a song about orange juice and nurtured a long desire to visit the republic of Tuva; land of the famed throat singers. One of the reasons inspiring this desire, he confessed, was to see a country whose capital city was named Kyzyl. Anywhere with a capital city of that name just *had* to be an interesting land! Alas, that wish was never fulfilled. During Feynman's lifetime, Tuva was part of the USSR and access by westerners proved to be far harder than solving some of the most difficult problems of physics.

Long before he knew of Tuva and probably before his first set of bongo drums, the juvenile Richard owned a toy wagon—a little red one, according to the story. One day, whilst playing with his wagon, he noticed something that most children of his age would simply have overlooked. There was a ball in the wagon and, young Richard observed, every time he moved the wagon forward, the ball rolled to the back. Then, when he stopped the wagon, the ball rolled to the front. Why, he wondered, did this happen?

Young Richard ran to ask that fount of family wisdom—his father. Now, many parents would probably have brushed aside yet another "Why?" question from their child. But the senior Feynman not only encouraged his son's enquiring mind but seems to have been largely responsible for turning Richard toward science in the first place. Yet, the answer that he gave the young Richard

FIGURE 2.1 Richard Feynman 1918–1988 (*Credit*: The Nobel Foundation)

was not as informative as the latter would have liked. The older Feynman explained that the ball behaved in this way because there was some property in balls and, indeed, in all material objects that caused them to remain in a constant state of rest or motion unless acted upon by an external force. In effect, what Richard saw was not so much the ball rolling forward or backward in the wagon as the wagon moving *under* a ball that tended to remain stationary.

The senior Feynman told his son,

> If you look from the side, you'll see that it's the back of the wagon that you're pulling against the ball, and the ball stands still. As a matter of fact, from the friction it starts to move forward a little bit in relation to the ground. It doesn't move back.

On hearing this, the younger Feynman later recalled that,

> I ran back to the little wagon and set the ball up again and pulled the wagon. Looking sidewise, I saw that indeed he was right. Relative to the sidewalk, it moved forward a bit.

> **Project 3: The Little Red Wagon**
>
> Try this for yourself. Either borrow your child's wagon (and let him/her know why. You might encourage another Richard Feynman that way!) or, if there is no toy wagon, find something else that will move accordingly. Place a ball, pencil or anything that rolls easily on the back and pull the "wagon" along for a short distance and then stop suddenly. What happens to the "ball"? Line up the ball (or pencil, or whatever) with something beyond the wagon and see whether it actually does roll back from its earlier position (relative to this external reference, not to the back of the wagon) or whether, as the young Feynman found, it actually "moves forward a bit".

The property of material objects responsible for this behavior, the senior Feynman informed his son, is called *inertia*. So far, so good, but then things took a downturn. We can guess what Richard's next question would have been and we can also guess his disappointment at being told "I don't know – nobody does!" or words to that effect. Why do material bodies possess inertia? They just DO! As Isaac Newton expressed it, inertia is an innate force existing within a parcel of matter, "a power of resisting by which every body, as much as in it lies, endeavors to preserve its present state, whether it be of rest or of moving uniformly forward in a straight line". Of the nature of this innate force, he did not speculate.

The tendency for a body to remain in a fixed position is experienced as that body's *mass*. Strictly speaking, we should call this the *inertial mass* of the body. It is measured by applying a force of known strength to the body in question and then measuring the resultant acceleration. This, by the way, is how astronauts measure their body mass whilst in the weightlessness of outer space. A normal pair of scales which we on Earth use to weigh ourselves would fail miserably in these conditions. Normal scales measure *gravitational mass* and can only function where the scales and the object being weighed are equally affected by the same gravitational field. Under normal circumstances on the ground, measurement of gravitational mass and inertial mass are the same, but that is not so for somebody floating around in an orbiting space capsule!

All of this appears too simple to concern us ... until we think about it a little longer. Then questions rise up like mushrooms after rain. Why *do* objects tend to remain in a fixed position? We are so familiar with this that we could go through life without giving it a passing thought. But is there any obvious *a priori* reason *why* bodies remain still rather than floating all over the place? And why should there be such a thing as mass? We can at least imagine a universe of mass-less, weightless, objects happily floating around like classical ghosts. We know that at least some particles are without mass (photons, for example). So why should the protons and neutrons and their ilk that make up material bodies also not be mass-less?

Those scientists who have taken these questions seriously have, over the years, put forward some ingenious hypotheses.

## Scientific Astrology?

OK, I admit that this subheading is an attention getter! But there is also a serious purpose behind it. Even if the positions of the stars and planets do not determine whether or not we will meet that elusive tall, dark stranger, may the fixed stars (or, more precisely, the remote masses of the universe) not influence us in some more subtle manner? The nineteenth century physicist and philosopher Ernst Mach, and those who came under his influence (one of whom was none other than the young Albert Einstein) thought so (Fig. 2.2).

A few moments ago, we mentioned how a body's inertial mass is measured (or, is manifested) by applying a force of known strength and then measuring the acceleration of the body in response to this force. What was then assumed, but unstated because it is self-evident, was that this measurement was against some reference; some measuring stick so to speak, which was itself part of a reference frame that did not share the acceleration of the body being measured. In other words, the inertial mass of the body was measured by accelerating it relative to an *inertial frame*. For the astronaut being weighed, the immediate frame was the space capsule, but even this was encompassed by the even

FIGURE 2.2 Ernst Mach 1838–1916 (*Credit*: H. F. Jutte)

larger inertial frame seen through the capsule's portals—the ultimate inertial frame of the distant stars and galaxies. For Mach and his colleagues, the remote masses of the universe constituted the ultimate inertial reference frame against which all inertial masses are manifest. In other words, *a body manifests inertial mass through being accelerated relative to the remote masses of the universe.* We can think about this in another way, namely, if there were no other masses in the universe (if there was just one body in an otherwise infinite vacuum), how could the inertial mass of that body ever manifest? There would be nothing to accelerate it and nothing to measure any acceleration even if this could occur. Indeed, the very concepts of rest and movement would then be incomprehensible. How could rest and motion be differentiated—how could they have any sense at all?—to a lone object in an otherwise empty universe?

In support of the Machian hypothesis, physicist Dennis Sciama calculated that the reaction force, experienced by an accelerating object, resulting from the combined gravity of all the matter within the observable universe is sufficient to explain that object's inertial mass.

Critics argue that if this is to be understood as a viable physical explanation for inertial mass, the gravitational effects of the remote masses would need to act instantaneously across the great gulf of intergalactic space. But it could be counter-argued that, since the universe was once very much smaller than it is now and that matter is not being continuously created, this criticism loses its force. The remote masses were not always so remote!

As an aside, we recall from the previous chapter that certain hypotheses were put forward to account for the cosmic redshift without involving the expansion of space and mention was made as to how a small number of astronomers were not very happy with the idea that the very high redshifts of quasars was cosmological in origin, suggesting instead that these were more likely compact objects ejected at high velocity from the nuclei of our own and/or other nearby galaxies. Although not mentioned in that chapter, one variation on the ejected-matter theme suggested that the redshift of the (supposedly) ejected quasars was not so much due to the velocity of these objects as to the possibility that they were composed of newly created matter, somehow produced and ejected by the nuclei of their companion galaxies. The speculation went like this. Because this newly-created matter had not had sufficient time to be influenced by the gravitational attraction of the remote masses of the universe, it possessed different inertial properties to the older objects comprising the bulk of the universe and this difference manifested in the lower energy (longer wavelength) of its emitted light. This hypothesis, together with the more common local-quasar version, is dead, buried and unlikely to rise again, but is worthy of at least brief mention with respect to the larger Machian issue.

## A Very Lively Vacuum

Another ingenious proposal for the solution of the inertia problem was put forward in the latter years of the previous century by Physicists B. Haisch, A. Reuda and H. Puthoff. Like Mach and those who followed him, this trio also looked to the universe at large for an ultimate inertial frame. However, for these physicists, the structure of that frame is not the macroscopic remote masses but the sub-microscopic vacuum fluctuations that pervade the whole of so-called "empty" space.

Somebody once called the hypothetical ether proposed by nineteenth-century physicists "that strange stuff that exists wherever there is nothing". Well, the ether, as initially conceived, does not exist. Nevertheless, there *is* a sort of ether in the form of a sea of virtual particles that (like its hypothetical predecessor) "exists wherever there is nothing". As briefly mentioned in the previous chapter, the theoretical reason for the presence of this sea lies with Heisenberg's Uncertainty Principle; one of the keystones of quantum theory. Essentially, this principle says that we cannot exactly measure certain attributes of a particle; we cannot simultaneously measure its position and momentum for example. However, for completely empty space or vacuum to exist, one *could* measure both the position and energy of any point in space. A point of vacuum (if truly empty) must by that very fact have zero energy; something which could be measured simultaneously with its position etc. in defiance of the Uncertainty Principle. It follows therefore that the energy of empty space can never be exactly specified. The vacuum is really a sea of constantly changing energy, manifested as virtual particles. Contrary to what might be thought, virtual does not mean unreal in this context. They are real enough, but come into being together with their corresponding anti-particle partners and immediately mutually annihilate. They are gone before they have a chance to truly manifest as real.

On the face of it, this sounds like far-fetched metaphysics, but in fact this sea of virtual particles has testable effects (for example, the Lamb shift in the spectrum of hydrogen, the Casimir effect manifesting as a slight attraction between plates brought into very close proximity and the irreducible electronic noise experienced in semiconductors) all of which have been verified to a high level of precision.

In the middle of the 1970s decade, research into this quantum vacuum by physicists William Unruh and Paul Davies suggested the presence of a curious theoretical phenomenon. If an observer moved through the quantum sea of virtual particles at constant velocity, the vacuum appeared the same in all directions, but if an observer accelerated through it, the vacuum takes on the appearance of a sea of low-level heat radiation. In practice, this radiation is too weak to measure, but its presence as a consequence of the theory intrigued both Haisch and Puthoff who independently noticed the similarity between acceleration/heat radiation and acceleration/inertial mass. Is there a hint here that accelerating through the vacuum could generate the resistance to acceleration that manifests itself as inertia?

Haisch began to develop his ideas in consultation with Rueda before learning of Puthoff's similar interest but, upon learning of the latter's work, the three joined forces to further develop their theory.

Briefly stated, Haisch et al. propose that the vacuum fluctuations produce a magnetic field "felt" by charged subatomic particles accelerating through it. These experience a Lorentz force (the well-known force that deflects charged particles moving through a magnetic field) which resists their motion. The larger the macroscopic body, the more particles it contains and therefore the stronger this resistance; manifested as greater inertial mass.

Even rest mass, the kind of mass that is equivalent to energy in accordance with Einstein's famous equation, might be explained in this way. According to Haisch, particles at rest are not actually still. They are in a constant state of jitters! And it is this jittering motion that gives them their rest mass. In the final analysis, rest mass and inertial mass are essentially the same thing; a very neat and convenient thesis indeed!

The reason why Haisch and his colleagues think that atomic particles are in a constant state of the jitters goes back to a proposal made by quantum physics pioneers L-V de Broglie and E. Schrodinger, about whom more will be said in a later chapter. These physicists were puzzled by a curious feature of electrons. They found that when low-energy photons are bounced off electrons, they scatter in a way that suggests the electrons are tiny balls of negative charge having a small but finite size. However, in interactions involving photons of very high energy, the electrons behave as if they are point-like charges. Schrodinger and de Broglie proposed that electrons really are point-like, but that this point jitters around within a finite volume. At high energies, the interaction is so fast that the electron appears frozen in place and its true point-like nature is revealed. At lower energies the interaction is slower and the jiggling motion of the electron causes the particle to act as if it were a fuzzy ball of finite dimension.

Haisch theorized that this jiggling motion is due to the buffeting of electrons by vacuum fluctuations and, in a mutually co-dependent chicken-and-egg relationship, also acts as the cause of these fluctuations. As the electron "jiggles" it emits electromagnetic radiation which fills the void with fluctuating radiation, manifesting in the creation of virtual particle/antiparticle pairs. The jiggling motion has affinities—albeit on a greatly reduced scale—to the Brownian motion of tiny material particles suspended in a gas or liquid medium. In the latter instance however, the buffeting is by the molecules of that medium. More will be said about this latter phenomenon later in this book. By the way, the name given by Schrodinger to the sub-microscopic counterpart of Brownian motion is *zitterbewegung*. One might wish that it had been called Schrodinger motion!

It might be worth mentioning here, simply in passing but holding Mach in mind, that for any point in space, the majority of electrons—and all atomic particles for that matter—within the region of the universe observable from that point, reside in the remote masses. As an aside, the co-dependent nature of particle mass and quantum vacuum fluctuations would seem to imply that the energetic vacuum could not exist prior to the material universe and that in the very distant future, if a time ever arrives when all material particles have receded so far from one another that, because of the expansion of space, they are effectively receding from each other at speeds greater than that of light, all vacuum fluctuations will cease, no particle will have mass and all there will be is ... *nothing*! Yet, this same mutual dependency also raises the prospect of manipulation of inertia and (who knows?) maybe even of the quantum vacuum itself. Although we have no idea how this could be accomplished at out level of knowledge and ability, this may change in the distant future. Maybe life itself will come to play a role on a grander scale than we can now imagine! (Strictly speaking, the above remarks are only dealing with the electromagnetic vacuum, however it is presumably correct to think that what applies to it applies generally and that all zero point—vacuum—energy can be similarly treated).

Returning to the present, it is interesting to note that Haisch's theory, by implying an energy-rich vacuum, also implies a cosmological constant contributing to the expansion of space at a cosmological level. This, as noted in the previous chapter, is just what observational cosmologists have found. However, the problem is that the cosmological constant derived from the theory of Haisch and colleagues is far, far, too large. It would have ripped the universe apart long ago! Maybe there is some as-yet-unknown factor that largely (though not *completely*) negates the cosmological constant, but one would require more than the mere wish to make the theory come out right to convince skeptics of the existence of such a factor.

That Cursed Elusive Particle!

If the discussion thus far teaches anything, it is surely that even the simplest phenomenon (a ball rolling backwards in a toy wagon for example), if pursued doggedly enough, uncovers layer upon layer of "weirdness" in the universe. What started out as a child's question has led us to the consideration of quantum vacuum fluctuations and the possible ways in which these may influence the macroscopic world.

Well, let's throw something else into the ring. The famous (infamous?) Higgs particle or Higgs boson. After a long search and at least one false alarm, it now seems that this elusive particle has been found, thanks to the efforts of the team at the Large Hadron Collider. Quite a deal has appeared in recent times in the popular press regarding this particle, provocatively called the "God particle" in many articles; a term which subconsciously hints to the reader that it is somehow prior to the very universe itself. Actually, physicists shun this moniker. The original expression was apparently the "god damned particle"; an expression having entirely different connotations! Presumably experimental physicists cursed its elusiveness as the quest for evidence of its existence came to take on the appearance of a wild goose chase.

The perceived need for a Higgs particle derived from attempts in the early 1960s by Peter Higgs and others to introduce inertial mass into the *Standard Model* of particle physics. For obvious reasons, mass must enter the scene somewhere, yet the equations of the SM simply go haywire once it *is* introduced. Clearly, something was missing in the model as it was then understood.

A breakthrough, at least at the theoretical level, was eventually made in 1964 by Higgs and his colleagues. Higgs et al. proposed the existence of a field permeating the whole of space that (simply expressed) exerts a resistance or drag on mass-less particles accelerating through it. It is this drag that manifests as inertial mass. So, to answer the young Richard Feynman's question, the ball rolls backward in the toy wagon because space is filled with a strange kind of field—now known as the Higgs field. The disturbance created by mass as it moves through the Higgs field is what gives rise to the Higgs particles, each of which has a very short lifetime.

Explaining inertia in terms of particles encountering a resisting "medium" that permeates space has a very familiar ring about it! It recalls the hypothesis put forward by Haisch and colleagues, except that in that scenario, virtual particles played the role that the Higgs field does in the alternative theory. Initially, the Haisch model was seen by some as a way of explaining inertial mass without the need to conjure up a new field and particle, but since the Higgs has apparently been discovered, does this mean that the alternate model must now be discarded?

Not really, according to Haisch and colleagues at the Calphysics Institute of Inertia Research. The Higgs field, they argue, only applies to the electro-weak sector of the SM and as such can account for just one percent or thereabouts of the mass of the protons and neutrons that constitute material objects. The Higgs process is, as they characterize it, "basically a transfer of energy from a field to a particle". In a *Scientific American* article of November 1986, M. J. G. Veldman uses a picturesque analogy which Haisch and colleagues quote as follows:

> The way particles are thought to acquire mass in their interactions with the Higgs field is somewhat analogous to the way pieces of blotting paper absorb ink. In such an analogy the pieces of paper represent individual particles and the ink represents energy, or mass. Just as pieces of paper of different sizes and thicknesses soak up varying amounts of ink, different particles 'soak up' varying amounts of energy or mass. The observed mass of a particle depends on the particle's 'energy absorbing' ability, and on the strength of the Higgs field in space.

This analogy enables us to understand (or, at least to picture) the process in familiar terms, but in Haisch's opinion it fails to tackle a deeper question which may be asked here, namely, why does a particle come to resist acceleration once it has soaked up energy from the Higgs field? Once again, we seem to be faced by the two alternatives of either understanding inertia as a fundamental property which cannot be explained any further *or* postulating that there is something about space itself that gives the Higgs field its teeth, so to speak. Haisch's suggested answer remains unchanged, Higgs field or no Higgs field!

Much more could be said about this topic, and no doubt much more will be said as the years pass. What the final answer—assuming that there will be one—turns out to be is for the future to know, or maybe not know as the case may be. Perhaps someday physicists will know the full answer to the question "What is inertia?" Perhaps a future juvenile Richard Feynman who asks his father why a ball rolls to the back of his toy wagon will be given the full explanation!

## Einstein's Universe

Ask anyone to name a genius, and there is every chance that they will say Albert Einstein. He has been cast as the archetypical genius, sage-like in his more advanced years with his shock of white and rather unruly hair. If someone is said to be "an Einstein" or "another Einstein", we know exactly what is meant. Einstein and genius have become synonymous terms in the minds of many. This characterization of him is fair enough, although it is undoubtedly true that others of equal and even greater mental capacity have breathed, and are still breathing, the air of this planet. The names Fred Hoyle and Stephen Hawking spring to mind from within the scientific community and, of course, science has no monopoly on genius. Einstein's IQ is estimated to have been around 160. Certainly genius level, yet by no means a record. Edward Witten, for instance, is said to top 180 in the IQ stakes, and even higher estimates are known. Yet, Einstein's claim to genius in the popular mind rests, it is fair to say, on his discovery that the universe is a lot weirder than anyone had previously believed.

FIGURE 2.3 Albert Einstein, 1879–1955, photographed in 1921 (*Credit*: Ferdinand Schmutzer)

Although he started from the notoriously "facts-only" philosophical position known as positivism (a philosophy of science championed by Ernst Mach, whose writings greatly influenced Einstein in his early years), his theories eventually brought to light so many counterintuitive and, frankly, non-common-sense aspects of the real world as to completely shatter the neat clockwork and comparatively easily understood universe in which scientists had come to believe by the turn of the last century (Fig. 2.3).

The positivist approach to physics tried to avoid all metaphysical questions. As already mentioned, and reminiscent of the chief character in a certain 1950s television crime drama, positivists wanted "only the facts; just the facts!" That was Einstein's approach when, in 1905, he wrote his epoch-making paper simply entitled *On the Electrodynamics of Moving Bodies*. Einstein began by defining space and time in very positivist terms; space is what we measure with a measuring rod and time is what we measure with a clock. The metaphysical questions of what space and time "really" are did not concern him. Armed with these definitions, he asked the question as to how the measurement of space and of time might change between two observers moving at constant velocity relative to each other. Imagine one of the observers, holding his clock and measuring rod, riding on a moving locomotive while his friend (armed with his very own clock and measuring rod) remained on the station platform. The rider of the locomotive measures the length of the window of his car, using the measuring rod that is riding with him and the observer on the platform does the same with his measuring rod as the locomotive rushes by. If asked whether the two measurements turn out identical, our immediate common sense answer is "Yes". Anything else seems unthinkable. Yet, Einstein was able to demonstrate that this answer is wrong. The person on the platform must see the window moving past him. This appears so obvious that it hardly needs mentioning, yet what does this seeing actually mean? Clearly, seeing is a shorthand way of saying that light from the moving window is entering the eyes of the observer on the platform and it is by means of this light that information about the length of the moving window is reaching him. Light is the carrier of information and, in this capacity, is vital for the measurement. It is, of course, no less vital for the locomotive passenger's measurement as well. If light plays such a vital role in both measurements, it follows that the measurements themselves will depend equally vitally upon light's properties.

Light moves through space at a finite—albeit very great—velocity. In round figures, it travels in vacuo at a speed of 186,000 miles (300,000 km) per *second*. Now that is fast, but it is not infinite. This much was known long before Einstein, having been demonstrated in 1676 by Ole Roemer (1644–1710) through observations of Jupiter's moon, Io. It was Einstein however who postulated that the velocity of light is an absolute constant. That is to say, its speed is always the same irrespective of the speed or direction of motion of the observer. Light from a source approaching an observer has the same velocity as that from a receding source. This makes it unique.

Einstein's other postulate appears at face value to contradict the above. This postulate states that it is impossible to determine absolute uniform motion. This, at least, agrees with common sense—up to a point. Uniform (non-accelerated) motion is essentially simply coasting along, and one cannot determine whether one is coasting or not without reference to something else; to the inertial reference frame in which one is located. Married together, these two postulates effectively see light as the ultimate reference frame. It is this marriage between the relativity of motion of all material objects and the absoluteness of the velocity of light which gave birth to the family of weird and apparently counter intuitive predictions of the theory that became known as *Special Relativity*.

Working from these postulates (which were actually given in the reverse order to that presented here—i.e. the impossibility of determining absolute uniform motion is the first postulate of Special Relativity and the absoluteness of the speed of light in vacuo is the second), Einstein mathematically deduced laws relating space and time measurements made by different observers moving uniformly relative to each other. Referring back to the example of the locomotive passenger and his friend on the station, Einstein's equations show that the latter measures the length of the window as being less than the former would measure it. Moreover, as the locomotive increases its speed, the station-bound person's measurement of the window gets shorter and shorter. If we imagine an especially advanced railway system where locomotives can approach the speed of light (let other considerations, both theoretical and practical, lapse for the moment!) Einstein's equations indicate that the measurement from the station reaches zero as the locomotive reaches light speed. All the while however, nothing changes for the passenger's measurements. This is not, let it immediately be said, a question of one or other of the observers making errors of measurement. The window *really does* get shorter as measured from the platform—whatever "really does" *really* means!

Similarly, if the observer on the station could see the clock on the speeding locomotive, he would notice that it ran more slowly than the one sharing the station with him. Once again, if the locomotive actually reached the speed of light itself, the on-board clock would stop *as perceived from the station*. For the locomotive passenger however, time would pass as usual.

> Project 4: Differing Reference Frames
> in Common Experience
>
> Although the really strange consequences of accelerating frames of reference do not make their presence felt until speeds approaching that of light are reached, some minor odd effects can be experienced at everyday velocities. For instance, have you ever been driving through the countryside, with the window of your vehicle down, and passed a roadside signpost? Have you noticed the swish as it moved past? But wait a moment! Was it not you in your vehicle that moved past it? But the swishing sound was what would be expected if you were stationary and the signpost was rushing past at high speed. Well, as far as you were concerned, that is exactly what happened. Relative to the framework of your vehicle, you were stationary while the signpost, trees and the rest of the world rushed by; making the right sound effects as they did!
>
> Take note of this next time you drive along an open road. It is a very simple observation of course, but it does form a tenuous link between our common experience and the far more profound issues raised by moving frames of reference in which relative velocities approach that of light itself.

## That Old Man Is My Son ... !

Of course, all this talk of locomotives moving so fast as to reveal these effects is purely hypothetical. In our normal experience, the effects about which Einstein was speaking are so slight as to be beyond measurement. However, they have actually been measured at high velocities.

Take the prediction of slowing time for example. Highly accurate atomic clocks carried around the Earth in jet aircraft have been found to register the so-called time-dilation effect to exactly the degree predicted by Special Relativity theory. In 1971, two American scientists, Joseph Hafele and Richard Keating, flew two atomic clocks around the world in two jet aircraft. One clock was flown in an easterly direction and the other in a westerly, with a third clock staying home for reference at the US Office of Naval Research in Washington. Compared with the Washington clock, the eastward-flying one was moving at the greatest velocity, being the sum of the aircraft's own speed together with that of the rotation of the Earth. From the same reference point however, the westward-flying clock was travelling more slowly. Flying against the rotation of the planet, its carrier was acting like the child who tried to run up a downward- moving escalator. According to the time-dilation effect, the eastward-flying clock should be slightly behind, and the westward-flying one slightly ahead, of the reference clock in Washington. The slight discrepancies in the time shown by these clocks were just as predicted by the theory. Then, in 2005, a similar experiment was carried out under the direction of the National Physical Laboratory in the UK, once again showing results that were accurate to within the acceptable error margin of the predictions.

In an even more dramatic demonstration of the reality of the phenomenon, time dilation is necessary to explain the strange history of mesons formed in the stratosphere through bombardment of cosmic rays. These particles have extremely short lifetimes—of the order of one millionth of a second—so that even travelling at the speed of light they cannot survive for more than a kilometer or thereabouts. Yet, amazingly, they have been detected at ground level; a distance some ten times greater. This seemingly impossible situation becomes explicable once it is appreciated that these particles travel close to the speed of light. If we could imagine that each meson carried with it a tiny clock that could nevertheless be observed by someone on the surface of Earth, Special Relativity tells us that this clock would run extremely slowly relative to our own wrist watch. In effect, these particles do carry a clock in the form of their rate of decay. It is the slow running of this clock, relative to a clock on the ground, that enables them to travel so far.

Conversely, a hypothetical observer riding the meson would see time running normally, but would experience a great contraction of space, so that for this observer, the meson's point of origin would be just 1 km above the ground, not 10 km as measured by an observer on the ground itself!

This effect has some potentially very strange consequences and if anyone doubts that we live in a weird universe, a few moments reflection on these and the nature of the universe that makes them possible will surely convince otherwise.

Imagine an astronaut of the distant future leaving Earth on a space ship capable of attaining relativistic velocities, i.e. velocities approaching that of light, such that the effects of Special Relativity become significant. We imagine the 30-year-old man saying goodbye to his 10-year-old son and setting out on a trip taking say, 4 years as measured by whatever types of clocks space ships of those days will carry on board. However, his destination is a planet of a star 40 light years distant. The space ship accelerates at 1G for a period of 1 year (again, as measured by on-board clocks) by which time it will be travelling very close to the speed of light. Another year of deceleration at 1G brings it to its destination (acceleration/deceleration at 1G means that all on board experience a very comfortable force equivalent to that of Earth's gravity). Business on the planet is quickly finished and the astronaut returns home in the same manner, a time (measured by the ship's clock of course) of 4 years having elapsed since he left home. But on landing, the (by then) 34-year-old astronaut is enthusiastically greeted by his 90-year-old son who has waited 80 years for his father's return!

The most familiar presentation of the time-dilation effect is given in terms of what has become known as the twin paradox. Essentially, this is the same as the father and son story, except that the space traveler and the stay-at-home relative are here assumed to be twins. When the astronaut left on his journey, both were of equal age but upon returning, the astronaut finds that his twin is now older than he. As it stands, although this situation is certainly counter-intuitive for a species that always believed in the constant flow of time, it hardly constitutes a paradox. So why is it known as the twin paradox?

The apparently paradoxical aspect derives from the relativity of motion. Motion, as we saw, is always relative to some reference frame. Thinking back to the locomotive passenger for example, it is clear that he is not in motion relative to the reference frame of the carriage in which he is riding, but is in motion from the aspect of the station. Yet, it is also true that the station is in motion with respect to the passenger and his locomotive carriage. Likewise, although a relativistic spaceship approaches the velocity of light as observed from Earth, it is equally correct to say that Earth approaches the velocity of light as observed from the perspective of the spaceship. Therefore one might think that, from the perspective of the astronaut, the twin remaining on Earth should be the one who ends up being the junior sibling at the former's return. Yet surely both cannot be right—not even in this weird universe!

... Or Is He?

An interesting comment on precisely this point was made by French Atomic Energy Commission physicist, Jean Charon. Charon pictured a relativistic spaceship of the distant future accelerating away from Earth at 1G for 1 year. At the end of that time, it would have attained a velocity very nearly that of light itself. But recall the example of mesons generated in Earth's stratosphere. These, as we saw, actually reach the planet's surface, even though their miniscule lifetimes should only allow them to travel about one tenth of that distance, even at the speed of light. This paradoxical situation, we said, arises because space itself has contracted from the viewpoint of the mesons themselves. An observer riding a meson would experience the distance between the particle's point of formation and the ground as less than 1 km, not the 10 km as perceived by an observer on the ground itself. This space contraction is the other side of the same coin as time dilation, and it is here that Charon's account becomes very interesting indeed.

At the point at which Charon's imaginary relativistic spaceship almost reaches light-speed, all forward space effectively contracts to near zero. At that point, the distance of the star Alpha Centauri at 4 light years from Earth, the Andromeda Galaxy at 2 million light years—even that of a distant quasar at 6 billion light years—all shrink to effectively zero. If the ship's captain is bound for Alpha Centauri, he just points his ship in that direction (it will probably be just a little more involved than this if it ever becomes reality) and decelerates at 1G. A year later, he arrives at Alpha Cen.

But if the Andromeda Galaxy is the desired destination, the spaceship is directed toward this object and, once again, after a year's deceleration at 1G, it arrives in Andromeda. The same for a remote quasar or for any other destination in the universe.

Charon then goes on to say how returning astronauts from galaxies millions of light years distant will have wonderful tales to tell their children. But wait a minute! By the time an astronaut returns from the Andromeda Galaxy, 4 million years will have passed on Earth. Their children will be ancient archaeology, not just ancient history! The Earth itself will be as alien as anything they may have encountered in Andromeda. Surely such a trip will be a one-way journey if or when it is ever made.

Not so, argues Charon. Clocks, including the biological variety, do indeed slow down on the spaceship, from the point of view of an observer on Earth. But, as the Earth itself speeds away from the spaceship at relativistic speed, as remarked earlier, an observer on the ship would equally perceive all clocks on the planet as slowing down and the inhabitants of Earth slowing in their aging process, relative to what was happening within the reference frame of the spaceship. But space also contracts for both observers just as time slows. The upshot of this has the same result as superluminal velocity, but without the violation of Special Relativity that this would involve.

Referring back once again to the example of stratospheric mesons, we see that from the point of view of a ground-based observer, the clock (in this instance, the particle's decay rate) carried by the meson runs slow, giving the particle extra time (from the point of view of an observer on the ground) to reach Earth's surface. It is as if the meson travels faster than light, although this is certainly not what is happening. Conversely, we may say that the distance travelled by the meson contracts, just like the window in our hypothetical very fast locomotive carriage. On the other hand, for a hypothetical observer riding the meson, the particle's clock (its rate of decay) does not change. Instead, as the Earth's surface accelerates toward the particle at close to the speed of light, clocks there run slow and the distance travelled by the surface of Earth as it rushes toward the meson, contracts. Whichever way we look at it, the meson travels a distance greater than its decay rate appears to allow, yet without exceeding the speed of light! It is by following this line of thought that Charon reaches his unconventional conclusion.

In contrast to Charon's argument, what might be called the "traditional" solution of the so-called twin paradox relies on a subtle asymmetry in the motion of the twins. The spaceship is not, of course, in a state of uniform motion. Clearly, it is accelerating (in Charon's example, at 1G). However, the members of the astronaut's family back on Earth are not experiencing acceleration. According to this argument, it is this asymmetry between the accelerated motion of the spaceship and the non-accelerated motion of the family staying behind that breaks the paradox. Alas, only the former experiences the benefits of time dilation.

Charon, nevertheless, will have none of it. He counter-argues that just because the astronaut in his account remained under an acceleration of 1G throughout his journey, he has constantly experienced the same force (1G) as his family back on Earth. "So what?" one might exclaim and, indeed, this exclamation would be justified if we only had Special Relativity to consider. However, 10 years after publishing his epoch-making paper detailing this theory, Einstein widened the Relativity vista into the theory of General Relativity in which both inertial and accelerated reference frames are included. Using this broader application of Relativity, he was able to demonstrate the impossibility of distinguishing between the effect of gravity and that of a non-uniform motion such as the acceleration of a spaceship. The realization of this principle of equivalence was recalled by Einstein as "the happiest thought of my life". Applying this principle, Charon correctly states that in his spaceship, the effect of the accelerated motion would be completely indistinguishable from the gravitational attraction experienced at the Earth's surface. *Both* the astronaut in the ship and his family back home therefore experience the same 1G. If the astronaut stepped onto a set of scales, these would register his weight just as they did back on Earth. A dropped stone would fall to the floor of the ship just as quickly as it would back home. So how, asks Charon, can the twin paradox be solved by speaking as if there are two types of acceleration;

as if the principle of equivalence failed to hold? Only, in his opinion, by reverting to the pre-Einsteinian myth of absolute space. He writes;

> In my opinion, there is nothing in Relativity to support an idea such as this [i.e. that a greater period of time has passed on Earth than that experienced by the astronaut]. The idea comes from a pre-relativistic method of thought; it is believed that the space in the universe is absolute, and that the Einsteinian revolution applies only to the notion of time. This misconception leads on to the idea that travelers moving with respect to each other at high velocities [approaching that of light] will age differentially. [*Cosmology*, p. 225]

On the contrary (Charon argues), as time and space are affected equally, the 4 years pass equally for both spaceship passengers and for the family back at home. Ironically, this has the appearance (superficially at least) of treating time, rather than space as an absolute. That is certainly not Charon's intention, but it is difficult not to understand it that way in so far as the symmetry of the time-dilation effect cancels itself out, giving time the appearance of an absolute. Moreover, if Charon's conclusion is correct, what is one to make of the Hafele/Keating experiment mentioned earlier—which, by the way, was performed the year after Charon's book was published? This is something about which the reader may like to ponder.

## The Weird World of Curved Space

As already been remarked, the theory of Special Relativity was not the end of Einstein's work. In 1916, he published his theory of General Relativity concerning both inertial and accelerated reference frames. We already saw how this shone light on the equivalence of acceleration and gravity. Also referred to as "Einstein's Theory of Gravity" General Relativity also enables gravity to be treated, not so much as a "force" in the way that Newton had assumed, but in a geometrical manner as the curvature of space itself. More precisely, it is the curvature of the space-time continuum as the concepts of space and time are no longer separable in terms of General Relativity. Although it is difficult for us to

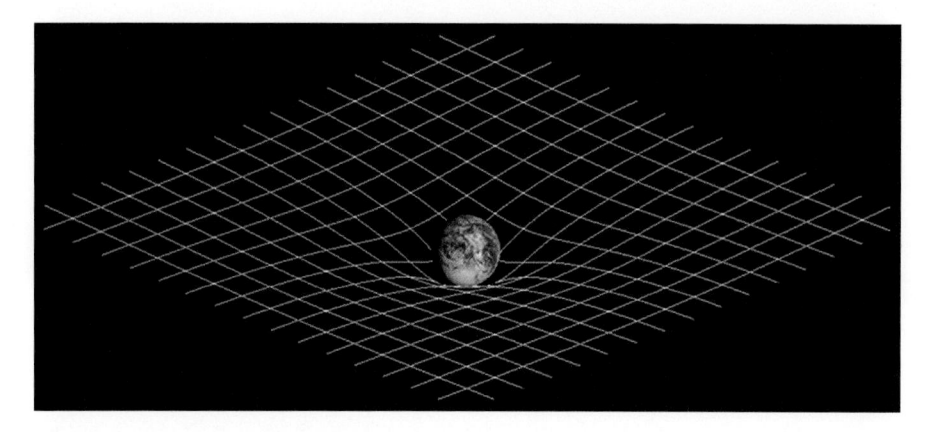

FIGURE 2.4 A two-dimensional representation of the three-dimensional warping of space in the presence of a parcel of mass (Johnstone, Earth image by Galileo spacecraft, credit: NASA)

understand how space, or space-time, can be "curved", Einstein demonstrated that this is indeed what takes place in the neighborhood of massive objects and it is because the path of an object traversing this warped region is directed inward toward the central body that something imitating a force of attraction toward that body is experienced. The path travelled through space-time is known as a geodesic and may be thought of as the counterpart of a straight line in flat or non-curved Euclidean space. A freely-falling or moving body follows a geodesic and light travels through space-time on a special geodesic known as the null (or zero) geodesic, effectively the geodesic of zero length. In flat (Euclidean) space, a geodesic is the familiar straight line, but when space is curved, the shortest distance between two points is also curved. A ray of light, passing through the curved space surrounding a massive object, bends in the vicinity of that object (Fig. 2.4).

Now, even according to Newtonian gravitational theory, a ray of light is predicted to bend a little as it passes close to a strong source of gravity. General Relativity, however, predicts a larger effect and this difference between the predictions potentially provides a good test of the two theories. But how could this test be carried out in practice? What experiment could be performed to distinguish whose prediction was correct? Just 3 years after Einstein presented his theory to the world, nature itself provided a splendid experiment.

Upon hearing Einstein's theory, British astronomer and Royal Society member, Arthur Eddington, noted that this aspect of the theory could be tested during a very suitable total solar eclipse which would take place on May 29, 1919. At the time this eclipse was predicted to take place, the Sun would be located in a region of the sky populated by several rather bright stars which, during the all-too-fleeting moments of totality, could be photographed close to the limb of the eclipsed Sun. Eddington figured that by comparing accurate measurements of the positions of these stars as photographed during the eclipse with those obtained at night—far from the position of the Sun—any displacement of their images due to light being affected by the gravity of the Sun should be apparent. The only problem was the track of totality; it would be a southern hemisphere spectacle only. Eddington, through the Royal Society, accordingly asked for the sum of 5,000 pounds to finance an expedition. Remarkably, despite wartime conditions at the time of his request, the finance was approved by the government. It has even been suggested that Eddington, a Quaker whose pacifist views were not welcomed by the government of the day, may have been given the request to get him out of the country for a while! Be that as it may, photographs of the eclipsed Sun and nearby stars were obtained from two sites south of the equator; from Sobral in Brazil and from the island of Principe off the coast of West Africa. Measurements from both sites agreed with one another and demonstrated the exact shift in position of the star images that General Relativity theory predicted. After 200 years, Newton's force of gravity had yielded to the curved spatio-temporal continuum of General Relativity!

Gravity and Time: Yet Another Time Dilation!

As if the slowing of time due to acceleration is not confusing enough, Relativity predicts that a similar effect takes place within a gravitational field; not surprising really, given the equivalence of gravity and acceleration. Briefly stated, clocks run more slowly (time passes less quickly) as gravity increases. Clocks on Earth's surface run a little slower than a clock being carried in a high-flying aircraft. We age more slowly if we stay on the ground, although the difference is not worth shunning air travel!

A moment's thought about this will show that this effect actually works against the time dilation effect discussed earlier in the context of the Hafele/Keating experiment. Indeed, in addition to the time dilation affected by acceleration, this gravitational time dilation (affecting the clock that remained on the ground) also needed to be factored into this experiment, once more with results that agreed with the predictions of Relativity to within acceptable range of observational error. So the experiment actually yielded strong support for both of Relativity's time-dilation effects.

The gravitational time-dilation effect differs from that resulting from acceleration, however, in avoiding the problems stemming from the apparent symmetry involved in accelerating reference frames.

Of course, gravitational fields such as that of Earth are so weak that the effect is negligible for all practical purposes, albeit not so slight as to escape measurement by highly accurate clocks as we have already seen. Yet, as we will see in the next section, there are places in the universe where gravity is so strong that all other forces bend before it; where this weakest known force of nature dominates the scene completely. In these regions, gravitational time dilation is far from insignificant.

But more of this a little later. For the moment, let's look at a little fantasy dreamed up by Thomas Gold, whom we already met during the previous chapter in his role as one of the three major proponents of the Steady State cosmology, and published in a letter to *Nature* in 1975. He called it the "Mother and baby paradox" although it does not really involve a true paradox thanks to the asymmetric nature of gravitational time dilation. Maybe it would be better called the "Peter Pan plan" in honor of the boy who stopped growing older.

Whatever we might like to call it, the story unfolds as follows. A mother desiring her child to remain young for as long as possible, each night collects masses from afar and, while the child is asleep, arranges them in a spherical shell around his bed (herself staying outside and at the completion of her task, retreating to some distance from the shell.) The next morning, she dismantles the shell and stores the masses far away from the child's bed. During the time that the child slept, the mother experienced, and aged through, the normal period of night (which Gold assumed to be 12 h) but the time for the child would have been shorter because of the extra gravitational attraction of the spherical shell. The child would have aged less than the mother!

It goes without saying of course that the masses required make this plan impractical, to put it very mildly. But the very fact that it can be presented even as a thought experiment without violating any physical laws is enough to demonstrate the weirdness that Relativity Theory has exposed in even the seemingly most commonplace phenomena of the universe.

## Gravitational Lenses, Black Holes and Cosmic Strings

Strange as it may seem, there are telescopes in the depth of space. No, I am not talking about alien observatories, but about entirely natural arrangements of matter that contrive to concentrate light from even more remote objects in a manner not too different from a giant lens. A little thought will show that this is not really as farfetched as it might sound on first hearing. After all, the bending of light rays around massive objects—as demonstrated so beautifully during the solar eclipse of 1919—is not vastly different from the bending of light as it passes through a refracting substance such as glass. The cause of the phenomenon *is* vastly different, (although that statement might need modification, as shall later be seen!) but the results are not. Following from this therefore, it is but a step to conceive of a situation where a sufficiently large mass, strategically placed, could so bend the light from a remote source as to actually produce a magnified image of that distant body, just as a glass lens can focus light from a distant object into an image of that object. It may seem counter intuitive (again!), but Relativity Theory shows that if a sufficiently massive object passes between an observer and a distant source of light, the warped space (gravitational field) surrounding the massive body focuses the light of the distant source and actually causes its apparent brightness to increase, not fade out as we might think, considering that what we are talking about here is in reality an eclipse of the distant light source! If the distant source is pointlike, multiple images of the "point" are formed around the eclipsing body. Actually, four images are created, arranged in a cross shape, although only in gravitational lenses of the highest resolution can all four images be observed separately. If the lensed object is an extended source such as a distant galaxy, distorted arc-like images are formed.

This phenomenon of gravitational lensing is now well established in astronomy. It was first observed in relation to quasars. Back in 1979, what appeared to be two identical quasars flanking a galaxy of far lower redshift were shown to be two images of the same quasar gravitationally lens by the mass of the intervening galaxy. Smaller scale instances of the phenomenon (termed gravitational microlensing) have been used since 1986 to detect planets orbiting stars in our own galaxy. That year, Princeton astronomer Bohdan Paczynski drew attention to the fact that individual stars should also act as gravitational lenses focusing the light of any background stars that they may eclipse. Because of the weaker gravity of the intervening mass (a star rather than an entire galaxy) the multiple images would remain too close together to distinguish with current technology (they would be separated by less than 0.001 s of arc), but the brightening of the "eclipsed" star should still be apparent and, if that star had any sufficiently massive planets orbiting it, secondary brightening caused by these should also be apparent under the right circumstances. Several planets have now been found using this method. Moreover, the arc-like images of very remote galaxies gravitationally lensed by the mass of intervening galaxy clusters have given astronomers a chance to observe objects very far away, whose light was emitted during a distant epoch of the universe. The true shape and dimensions of these objects have been deduced from their distorted images, allowing information to be gleaned about galaxies that, without the benefit of gravitational lenses, would remain beyond the ken of contemporary telescopes. All in all therefore, this exotic consequence of Relativity has been put to some very good employ by observational astronomers in recent decades.

Ironically, although Einstein himself may have noted the possibility of gravitational lensing as early as 1912, he dismissed it as being of no importance. He seems to have believed that, even though it occurred as a consequence of the theory he discovered, it would remain unobservable and would never be of any practical value. The person who really put the phenomenon on the map, so to speak, was not Einstein but an amateur physicist and engineer by the name of Rudi Mandl. A brief account of Mandl's pestering of Einstein has already been given in my *Weird Astronomy*, but it is worth recounting here as his contribution to our awareness of this important consequence of General Relativity can scarcely be over stated. Mandl wrote to Einstein on several occasions arguing for the existence of the phenomenon, but it seems that Einstein considered this "amateur" to be little more than a crackpot and procrastinated over Mandl's suggestion of publication. Eventually, Mandl journeyed to Princeton and put the case for gravitational lensing to Einstein in person. Einstein (grudgingly it would seem) eventually gave in and published a short item on the subject in 1936, more to get Mandl off his back than to draw the attention of the scientific community to an important natural phenomenon. This last point is not just conjecture. In a letter to *Science* shortly after the publication of his paper, Einstein wrote that the conclusion was a useless one that he only made known because "Mister Mandl squeezed [it] out of me" and that he published the result only because "it makes the poor guy happy". Admittedly, some of Mandl's ideas did get a bit farfetched (such as gravitational lensing of cosmic radiation causing biological mutations and mass extinctions), but he surely deserves the credit both for his independent discovery of the phenomenon and for being the moving force behind its wider recognition (Fig. 2.5).

Incidentally, the example of gravitational lensing having the highest resolution is that of four images of a single quasar flanking an intervening galaxy and widely known as the *Einstein Cross* or, more formally, as G2237+0305. In *Weird Astronomy*, I noted that this is a rather ironic title, considering Einstein's skepticism about the phenomenon, not to mention his arrogant treatment of Mandl over the issue. I remarked that the object would more appropriately be re-named the Mandl Cross, although I fully realize that the former name has become too familiar to be changed now. However—and this is a serious suggestion which I earnestly hope will be taken up by others—I now propose that the arc-like images of distant galaxies caused by the gravitational lensing of light passing through intervening galaxy clusters, be at least semi-officially known as Mandl Arcs. Surely that is not too much to ask on behalf of an insightful amateur scientist who was treated so dismissively by the famous genius! (Figs. 2.6 and 2.7)

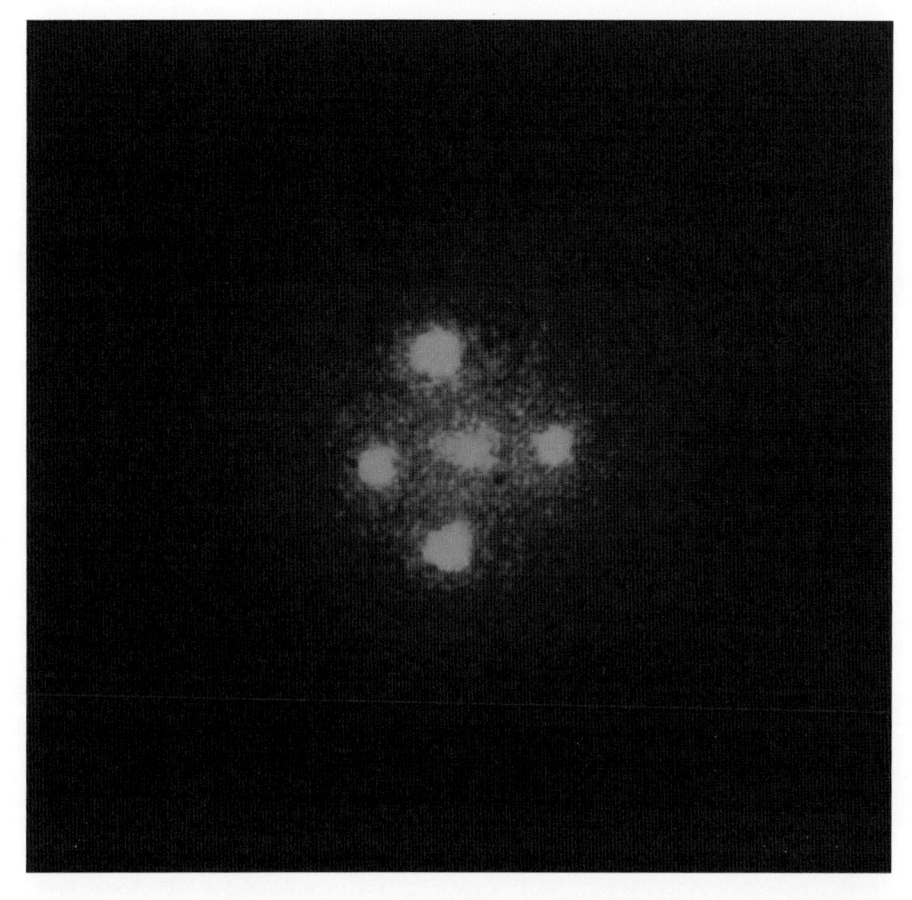

FIGURE 2.5 The "Einstein Cross"; four images of a single distant quasar resulting from the gravitational lensing of a foreground galaxy (shown here as the fuzzy object in the middle of the "cross") (Faint Object Camera, Hubble Space Telescope. *Credit*: NASA/ESA)

Now, gravitational lensing may seem strange enough. But it is nothing compared to another weird consequence of General Relativity—the black hole.

Nearly everyone has heard of black holes ... and I don't mean the infamous one of Calcutta! As used in astronomy, a black hole is a region of space where gravity is so strong that light itself cannot escape. From one point of view, this can be understood in a quite straightforwardly Newtonian manner. We know that a rocket must reach a certain threshold velocity (the escape velocity) to leave Earth and venture into outer space. Those of us old

FIGURE 2.6 Galaxy Cluster Abell 2744 showing gravitationally lensed images of background galaxies [*Credit*: NASA, ESA, J. Lotz, M. Mountain, K. Koekemoer & HFF Team (STScI)]

enough to remember the Apollo Moon landings will also recall how much less thrust was required to lift the lunar modules from the Moon's surface to rejoin the command modules and return to Earth. The escape velocity of the Moon (thanks to our satellite's weaker gravity) is a lot less than that of Earth. Conversely, for an object more massive than Earth (Jupiter or the Sun for example), escape velocity is higher. Imagine an object sufficiently massive,

FIGURE 2.7 A spectacular array of "Mandl Arcs" (highly distorted gravitationally lensed images of background galaxies) festoon this image of galaxy cluster Abell 1689 (*Credit*: NASA, N. Benitez (JHU). T Broadhurst (RACAH Institute of Physics/The Hebrew University), H. Ford (JHU), M. Clampin (STScI), G. Illingworth (UCO/Lick Observatory), the ACS Science Team & ESA)

and we can see that its escape velocity will come to equal or even exceed that of light itself.

The first person to appreciate this (or, at least, to publicly express it) was English clergyman, scientist and philosopher John Michell (1724–1793), who calculated that an object sharing the Sun's density but having 500 times its radius would possess such a strong gravitational field as to prevent the escape of light and manifest as an absolutely black object, absorbing light, but unable to emit it. He referred to such an object as a "dark star". Similarly, the French scientist P. Laplace reached a similar conclusion, albeit differing in the estimate of the size of such an entity. Modern computations show that the English clergyman was nearer the truth, concerning the computed mass though not the dimensions of these bodies, than the celebrated French scientist.

Both of these black hole pioneers employed Newtonian theory of course and it is unlikely that either would have been tempted to coin the term "black hole" for their hypothetical objects. There is no reason for thinking that either believed such entities to exist in the real universe, even though there seemed no theoretical barrier to their presence. The term "black hole" itself appears to have been first used by journalist Ann Ewing in 1964, but became accepted jargon following John Wheeler's use of the expression 3 years later. However, between Michel/Laplace and Ewing/Wheeler, there had been a very big change in the way in which we think of such things. The passage from "dark star"(with its connotations of an enormous solid object) to "black hole"(conjuring up thoughts of a super vacuum of some type) can only be understood by a fundamental change in the way we think of gravity and, indeed, of space itself. This change was the product of General Relativity. In short, once we come to picture a gravitational field around a material body, not as a region of space affected by a force tugging on all other objects as if the body is trying to capture them but as a sort of warp in space itself, the scene is set for belief in a hole in space. Science fiction loves it, conjuring up all sorts of tales about spaceships passing through these strange portals into alternative universes and such like.

In reality, black holes are no less weird, but before looking at this weirdness, it should be mentioned that there is another important difference between the black hole of Wheeler and the dark star of Michel and Laplace. Michel, as we saw, thought of his hypothetical dark stars as having densities similar to that of the Sun, albeit of far greater mass and, therefore, dimension. In the real universe as we now understand it, such a thing could not exist. Black holes are actually collapsed objects with dimensions far smaller than their total mass would suggest. The classic example is the stellar-mass black hole or collapsar. These are the corpses of stars which, during their lifetimes, were truly massive objects. The mass of a star in life critically determines what sort of stellar corpse it will become in death. For stars having masses below one half that of the Sun, life is extremely long and its end relatively sedate. Astrophysical theory indicates that all but the very smallest stars (those star/planet "missing" links known as brown dwarfs) will become brighter toward the end of their lives—the astrophysical version of the old adage of a candle burning brightest just before the flame goes out—but the end itself will be a simple fading into darkness. Their end will be as dark stars, albeit very different from those envisioned by Michel and Laplace. Black dwarfs is the better term for these stellar corpses. Small stars take life so easily that their life expectancies are greater than the present age of the universe, so there are no black dwarf stellar corpses at this time in the cosmic calendar. Actually, stars at the upper mass limit of this class are thought to go through the dense and hot phase of helium-burning white dwarfs first, but none has reached even this stage at the present cosmic epoch. This extra complication need not concern us here.

Stars of the next size bracket—from around half the mass of the Sun to some ten times more massive, have shorter life expectancies, living for several billions of years, and end a trifle more dramatically. Eventually running out of the hydrogen necessary to fuel their internal thermonuclear furnace, stars of this class eventually experience the collapse of their cores into a form of luminous stellar corpses known as white dwarfs, in the process puffing their outer layers into surrounding space as large and increasingly diffuse clouds called planetary nebula; this rather inappropriate title coming from the disk-like appearance that some of their number assume in the eyepieces of small telescopes. White dwarfs are amazing objects—planet sized but of stellar mass and density so great as to scarcely be imagined. To take just one (albeit somewhat extreme) example, the white dwarf known as Kuiper's Star weighs in at one solar mass, yet has a diameter approximately equal to that of the planet Mars. A thimble full of its constituent matter, if it could somehow be weighed at the surface of the star itself, would top the scales at several thousand million tons. So compressed is everything about this star that it has been estimated that its atmosphere is probably less than 20 ft (about 4 or 5 m on most estimates) deep (Fig. 2.8).

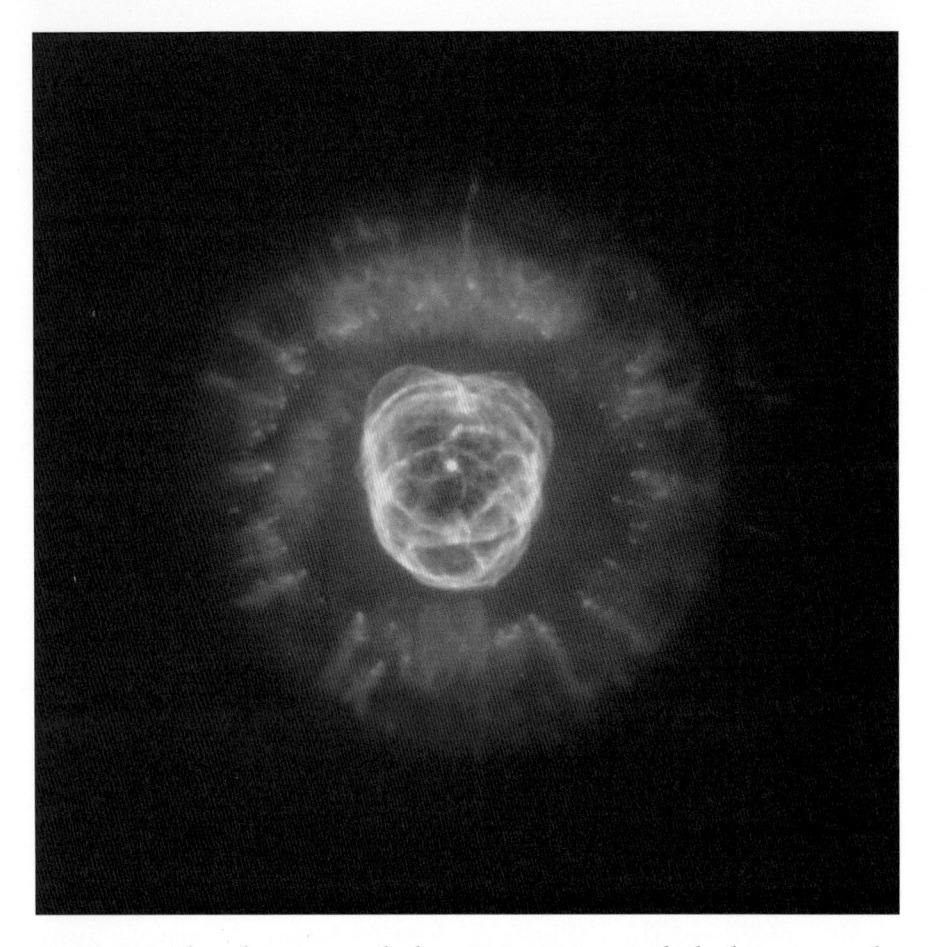

FIGURE 2.8 The planetary nebula NGC 2392, popularly known as the "Eskimo Nebula" [*Credit*: NASA, ESA, Andrew Fruchter (STScI) & the ERO Team (STScI + ST-ECF)]

Although white dwarfs start out far hotter than the surface of the Sun, they too cool slowly over time and will eventually become super dense black dwarfs. But once again, we will need to wait until the universe is quite a bit older before we have any examples of these.

Larger stars—those really massive brutes of 10–20 solar masses that blaze brilliantly across the light years—have lifespans measured in millions rather in billions of years and go out in a true blaze of glory. In common with those of roughly solar mass, these also exhaust their fuel (though in only a moment cosmologically speaking) and suffer core collapse. However, in these instances the collapsing core is so massive that it crushes its constituent atoms even beyond the stage of the white dwarfs. The collapse of a stellar core of white-dwarf mass is halted when the electron shells of the core's atoms press hard against each other. But when this mass is exceeded, even the electron shells themselves are forced into their nuclei, resulting in an exotic substance known as neutron matter. In effect, the core becomes a giant neutron of incredible density. Because most of an atom consists of the space between nucleus and electron shells, it can be appreciated that when the shells are driven into the nucleus, the size deflates catastrophically. This in turn means that a sphere of "ordinary" matter (composed of atoms having their nuclei and shells unchanged) the size of a planet can be squashed into something that could sit within the central region of a medium sized town, although its powerful gravitational field would do terrible things to the surrounding countryside! The collapse of a stellar core into an object of this nature triggers a horrific rebound that sets off thermonuclear fusion within the remaining body of the dying star before blowing the entire thing to pieces in the spectacular fireball of a core-collapse supernova. After the fireball (which may equal the combined light of a moderately large galaxy) fades away, the core remains as a rapidly spinning neutron star or pulsar. The radiation emitted by this object will, for a time, excite the nebulous remnant of the once mighty star to glow with a pale milky light; a well-known example of such being the Crab Nebula in the constellation of Taurus, more formally known as M1 and long identified as the gaseous remnant of a supernova seen in the year 1054 (Fig. 2.9).

Nevertheless, even those giants whose cores end up as neutron stars are not the very largest of all stars. A comparatively small number of behemoths having masses exceeding that of 20 Suns—even rare goliaths of around 100 solar masses—are known, and unless they can shed much of their mass in the meantime, when these objects run out of fuel and come to the end of their transitory lives not even the pressure of an atomic nucleus can halt the gravitational collapse. Stars that weigh in at between 25 and 90 Suns have cores in the range of 5–15 solar masses. According to (Relativity) theory, stellar cores with masses of this order collapse all the way to geometric points of infinite density;

FIGURE 2.9 The Crab Nebula M1 in Taurus (*Credit*: NASA, ESA & STScI)

in other words to singularities. Now, it is taken as an act of faith that nature abhors singularities at least as much as it abhors a vacuum and as such the prediction of these things is taken as good evidence that Relativity Theory must break down at very small dimensions. Presumably, Quantum Theory, about which more will soon be said, will save the day by somehow avoiding the singularity. Also assumed as an act of faith is the belief that the universe forms a consistent whole in the sense that a theory describing the macroscopic (such as General Relativity) and one describing the microscopic (Quantum Theory, for instance) are not so radically diverse that the twain will never meet. On the contrary, it is assumed that both theories really do describe the same united universe and will one day be married into a single all-encompassing theory covering in a single vast sweep the very large and the very small. That day, however, has not yet arrived and so we still treat these ultimate collapses as if they proceed all the way to singularities.

In any event, it is somewhat ironic that the very largest of these giants go out not with the bang of a great supernova, but with a whimper. Stars having cores greater than 15 solar masses collapse without any supernova marking their demise. Those having cores from about 5 to 15 solar masses do end in supernova fireworks, but the nature of extent of these displays depends upon whether the dying star was a slow or a rapid rotator during its lifetime. If the former, the supernova is faint by the standard of such events. However, the situation is quite the contrary for rapidly spinning stars of similar mass. High-energy jets shoot out of these collapsing giants, lighting up the surrounding shell that was once the body of the star, in a supernova event of such brilliance that some astronomers have awarded it the title of hypernova. This term is, however, not restricted to these objects and its use can be confusing. Whatever they are called though, the initiation of these mighty stellar explosions is also accompanied by a brief burst of gamma rays.

Whatever *really* forms following the collapse of the most massive stars, we know that the space-time continuum is so warped (i.e. that the "gravitational field" is so strong) in the immediate vicinity of the collapsed object that a region of space exists that acts like the dark stars of Michel and Laplace. Light may enter, but it cannot escape. This region of space is spherical and surprisingly small. A black hole of three solar masses spans just 11 miles or 18 km. Even one as massive as 100 Suns would have a diameter of only 370 miles (600 km) or so. The "surface" of the sphere—that is to say, the distance from the center at which something would need to travel at light speed to escape, is known as the *Schwarschild radius* in honor of K. Schwarschild who discovered the possible existence of such a region back in 1917. Anything within this region is, in effect, cut off from the rest of the universe. A sort of absolute horizon exists at the Schwarschild radius, over which light and material objects travelling toward the collapsed star can cross, but beyond which nothing (not even light itself) can return.

This is known as the event horizon, and the spherical region of no-return that lies beyond it may truly as well as picturesquely be called a black hole. Incidentally, it is generally held that a black hole is fully characterized by an external observer by means of just three parameters, namely mass, electric charge and angular momentum. All other information characterizing the collapsed star that formed it, or any other object that was unlucky enough to fall into it, disappears behind the event horizon. For some reason, this other information is metaphorically spoken of as hair, so the statement that all of this vanishes is often referred to as the no hair theorem. Mathematicians, however, tend to balk at this use of theorem and would rather speak of the no hair conjecture; a better term really, as a conjecture lacks the rigor of a true mathematical theorem.

Massive collapsed stars are not, however, the only possible sources of black holes. The possibility has been raised that black holes of all sizes and masses may have been created during the Big Bang itself. There is no evidence suggesting this and the inflationary scenario would seem to work against it, but it remains theoretically possible that the Big Bang or some other extreme event about which we currently know nothing could be capable of spawning black holes far smaller than any collapsed star. Any parcel of matter, if compressed sufficiently, will collapse into a black hole. The only problem is achieving sufficient energy to collapse it. But given this, truly tiny black holes could exist. The Schwarschild radius of a black hole having Earth's mass is just one centimeter. Hypothetical asteroid-mass black holes go down to sub-atomic dimensions.

On the other end of the scale, truly gargantuan agglomerations of matter can make huge black holes which, moreover, do not require high densities and thereby avoid the contentious issue of singularities. If the stellar population of a galaxy of about 100 billion stars could be packed into a region of space having a diameter of approximately 100 times that of the Solar System, the mutual gravity of these stars (each of which would still have a little surrounding room to move) would be such that light could not escape from that region. It would be a very large black hole. Massive black holes (albeit having higher densities than the hypothetical one of this example) are known to reside at the cores of large galaxies. Our own Milky Way sports a million solar-mass example believed to be some 3.7 million miles (6 million kilometers) across. By galactic standards however, even this is rather small. There is now evidence that ultra-massive black holes weighing in at between 10 and 40 *billion* solar masses inhabit the cores of very large galaxies such as those at the centers of galaxy clusters. Yet, there is one even larger black hole with which we are all acquainted. It is called the universe.

This may seem a very strange way of thinking about it, but the entire visible universe may be regarded as a black hole in the sense that the gravitational attraction of the total mass within it is so great that not even light can escape. The escape velocity, so to speak, of the universe exceeds the velocity of light. Unlike all other black holes, we are seeing this one from the inside as it were. The event horizon is not that which hides the inside of the black hole from an observer, but is the edge of the sphere of observation for any observer within the universe. So if we care to know what it is like inside one black hole, just look around and see!

We spoke in the previous section about the odd phenomenon of gravitational time dilation. We saw that although the gravitational well of Earth is sufficient to yield a time dilation measurable by the most accurate of clocks, it is too small to be of any practical consideration. But when we come to the gravitational fields of objects such as neutron stars and, especially, black holes, the situation is very different. Imagine an observer located at some safe distance from a black hole watching a spaceship flying toward this object. The observer is wearing a watch which, we imagine, marks each second of passing time by emitting an audible beep. A similar device is worn by the captain of the spaceship. In addition to providing audible beeps for the benefit of the owner of each watch, there is also an accompanying radio transmission such that the beeps of the space captain's watch are also heard by the external observer and vice versa. When the spaceship is distant from the black hole, the two sets of beeps will be equal (we will overlook the delay due to the finite velocity of the radio waves), but as the spaceship approaches the event horizon of the black hole, a curious phenomenon will be noticed by the external observer. The beeps marking the seconds get increasingly longer. They also decrease in tone (become deeper), but we will leave that aside for the present. For the space captain however, both the beeps from his watch and those from the observer will remain unchanged. (The perceptive reader will no doubt note at this point that the spaceship should also be accelerating under the increasing gravity of the black hole and that the time dilation due to acceleration should be apparent. The beauty of a thought experiment however, is that such problems that would beset a real physical experiment can be avoided by any fantastic method, just so long as it does not violate the physics being demonstrated. In this case, we might imagine that the spaceship has some sort of thrusters acting opposite to the gravity of the black hole in such a way that the ship's velocity remains constant even as it nears the event horizon!)

As the spaceship approaches the event horizon of the black hole, the ever-lengthening beeps mark the every-increasing length of the second, relative to the passage of time experienced by the external observer until, as it crosses the event horizon, the bleep becomes a continuous signal. The final second lasts forever as the spaceship's time stops relative to the external observer. For the spaceship's captain however, the seconds pass at their normal rate—all too quickly in fact, as the crossing of the event horizon now cuts him off forever from the rest of the universe. If the black hole he enters is a stellar mass one, he at least has little time to regret his loss as the steep gravitational gradient pulls the spaceship and its captain into a narrow ribbon before tearing both to pieces. If the hole is a very massive one he would at least have longer to contemplate his fate.

This cessation of time at the event horizon, as measured by an external observer, is sometimes represented as implying that ghost images of everything that has ever fallen into the black hole will remain visible around the periphery of the event horizon; a sort of everlasting diorama of things past. One might indeed expect the surface of the collapsing star to remain visible, frozen in time during the final stage of its implosion.

This, however, neglects another feature of gravity, namely, the stretching of electromagnetic waves in a powerful gravitational field. Light and other forms of electromagnetic radiation cannot be slowed by passing through a gravitational field, but energy is nevertheless lost and this loss is manifested by the lengthening (stretching out, in effect) of the wavelength of the radiation. This, we might recall from the previous chapter, was once put forward as at least a partial explanation for the very large redshifts of quasars. It was argued that if quasars are very massive objects of relatively small diameter, the resulting very powerful gravitational field associated with them should cause a significant stretching of the light waves emitting from them and that this would be observed as a strong redshift. This is correct physics, even though other observations of quasars indicated that the region from which their light emanated could not be dense enough for the gravitational redshift to be applicable. Gravitational redshift applies to radio waves as well as light of course, hence the earlier remark that the beeps marking passing seconds in our thought experiment should decrease in tone as their source plunged deeper into the black hole's gravitational field.

Taken to the ultimate, any electromagnetic radiation being emitted by a source plunging into a black hole will reach infinite wavelengths and zero intensity as that source crosses the event horizon, thus neatly avoiding an eternal gallery of ghosts clustered around the hole.

It should also be mentioned here that if the astronaut plunging toward the black hole somehow manages to turn his spaceship around at the last moment and return from the brink of the event horizon, he will find that he has aged less than the external observer. If the external observer was his twin, the astronaut will now find that he (the external observer) has become his older sibling.

In theory, it is even possible for the astronaut to use his approach to the brink of the black hole as a sort of time machine into the future. Thus, if he is by some means able to bring his spaceship into orbit almost astride the event horizon but without actually crossing it and disappearing into the hole, time within the spaceship will almost cease, but only as measured by an observer well away from the black hole. Then, after a certain period (as measured by a clock within the spacecraft) if the astronaut can pull away from the black hole, he will find himself in what he would consider to be the future. That is to say, while only a short period passed for him and for his spaceship, far longer stretches of time went by in the universe outside of his restricted environment.

It would be a pity however if he did not like the future in which he found himself. Such a "time trip" at the brink of a black hole is a one-way journey only!

Not-So-Black Holes?!

In the latter years of the 1970s, black hole enthusiasts were given an unexpected surprise courtesy of brilliant English physicist Stephen Hawking. On a purely theoretical basis, Hawking discovered that black holes may not be quite as black as had hitherto been believed. On the contrary, they should be sources of weak radiation and actually possess both a very low temperature and a very low luminosity. This temperature and luminosity, in theory, adds a little hair to the classic three properties of mass, angular momentum and electric charge which alone characterize black holes according to the no-hair conjecture. Weak though this radiation is however, because of the equivalence of energy and mass, the very fact that black holes are capable of radiating energy means that they are also (contrary to everything previously believed concerning them) capable of losing mass.

How is any of this possible? How can an object having a gravitational field of such magnitude as to trap radiation itself be capable of losing mass by emitting radiation? It does not seem to make sense.

Nevertheless, there is a way of interpreting this process through the weird world of quantum physics. More will be said about this in a future chapter, but already we have had cause to allude to Heisenberg's Uncertainty Principle and its implication of a constant flux of energy in the quantum vacuum. We simply recall here that this restless quantum sea at least in part takes the form of particle/antiparticle pairs (essentially electrons and positrons) popping into existence and instantaneously annihilating one another. It is well known that a particle and its antiparticle annihilate on contact, but the pairs that are constantly coming into existence throughout the whole of space effectively cancel each other out of the equation instantaneously. For this reason, as already mentioned, they are referred to as virtual particles, where "virtual" is being used in a rather special sense that does not equate with "unreal" or "hypothetical". Although the thought of an endless sea of particles popping into being and immediately ceasing to exist sounds like some totally hypothetical formulation, it does have verifiable (and falsifiable) predicted effects, which have not only been observed but measured to a high degree of precision and found to be in remarkable agreement with the theory's predictions. As briefly mentioned earlier in the present chapter, the two principal effects are the Lamb Shift and the Casimir Effect. The first of these predicts that the electron shell of hydrogen atoms will be perturbed by the presence of virtual particles and that this will be revealed in a slight shift in the spectrum of hydrogen. This effect was discovered by experimental physicist Willis Lamb and its magnitude found to agree with theoretical prediction to a high degree of precision.

The second consequence of these particle/antiparticle vacuum fluctuations was predicted by Hendrick Casimir in 1948. Casimir predicted that an apparent force of attraction should exist between two close perfectly conducting flat plates. He reasoned that the plates should essentially cut off the space between them from the surrounding sea of seething vacuum fluctuations. In effect, the surrounding vacuum will push the plates closer together. Casimir was able to calculate the strength of this apparent attraction and, around half a century later when it was finally observed, the measured strength agreed with calculations to within an accuracy of nine decimal places.

These effects, and their precise agreement with theoretical predictions, convince most people that the virtual particles are real and not just a convenient way of thinking about the quantum vacuum. But if real, they must also be susceptible to the strong gravity of a black hole and we may ask what happens to those virtual particle/antiparticle pairs that come into existence right at the event horizon of one of these objects. It is here that black hole radiation (now known as Hawking radiation in honor of its discoverer) is given physical interpretation. Briefly stated, one member of the particle/antiparticle pair crosses over the event horizon into the black hole. In so doing, it effectively vanishes from the rest of the universe, leaving the other member of the pair as a widowed particle. The disappearance of the partner particle over the event horizon saves the widowed particle from annihilation. From being a virtual particle, it now becomes a real particle and escapes the clutches of the black hole as Hawking radiation. As it does, it takes away an infinitesimal portion of the mass of the black hole!

Hawking's calculations demonstrate that as a black hole loses mass via this process of evaporation, it will not only shrink but will also increase the intensity of radiation that it emits. In other words, the smaller it becomes, the faster it shrinks and the hotter and more luminous it grows until, finally, when it reaches a size smaller than that of an atomic nucleus, it explodes in a burst of gamma rays. The black hole briefly becomes a "white hole"!

As far as we know, the smallest black holes are those of stellar mass. We are unaware of any method capable of forming smaller ones since the first few instants of the Big Bang. But stellar black holes radiate so feebly that they will not reach the explosive stage until they are billions of billions of times older than the present age of the universe. The lifetime of a 30 solar mass black hole, for example, is estimated as a whopping $10^{61}$ years. Moreover, the feeble loss of mass through Hawking radiation would be more than offset by energy and matter falling into the hole from surrounding space. If black holes of stellar mass really are the smallest that can form, the issue of complete evaporation and explosion of these objects is something about we need not concern ourselves for a long time yet!

However, the situation is different if small black holes were generated during the Big Bang. Those having masses equivalent to small asteroids should be exploding right now. Gamma ray bursts have indeed been observed, but none has had the characteristics expected for a black hole explosion. All have been too far away, for one thing. Black hole explosions (though not something that we would want happening too close to Earth) are not powerful enough to be detected at distances greater than the nearest stars, so although this rules out the observed gamma ray bursts, it also means that absence of evidence is not in itself strong evidence of absence. Nevertheless, it is quite probable that small black holes were not produced in the Big Bang and that the stellar mass objects truly do represent the minimum size of actual objects of this type.

Let us now look at a really tricky question! We already mentioned that the observable universe is a sort of black hole. From our point of view inside the hole, the event horizon presents us not with a barrier between the inside of a black hole and the rest of the universe but as a barrier between the universe that lies within our possible range of observation and the universe outside of this sphere. The event horizon of the universe—otherwise known as the de Sitter horizon—is the place where expanding space recedes from the place of observation at the speed of light. Now, it would appear to follow that if Hawking radiation is generated at the event horizons of black holes found within the universe, the same process should operate to generate this radiation at the event horizon of the black hole that defines the observable universe itself. This indeed is believed to be the case. The de Sitter horizon should theoretically emit very weak Hawking radiation, or de Sitter radiation as it is alternatively known in this instance.

But this is where things get even more complicated than they are already. The event horizon of an ordinary black hole at least has the decency to remain still. It has its fixed position in space and this remains unchanged irrespective of where the observer of the black hole is located. However, this is clearly not so for the de Sitter horizon. As the reader will readily appreciate from Chap. 1, the edge of the observable universe is, from a spatial viewpoint, dependent upon the location of the observer. Like a rainbow, it moves as the observer moves and, just as no two people actually see the same rainbow, so no two observers see exactly the same de Sitter horizon. In this respect, the event horizon of the observable universe has more in common with the familiar horizon we see on Earth than with that of a stellar black hole, but it is this very familiarity that breeds, not contempt in this instance but a head-banging difficulty.

Take two points in space, A and B. Point A is your position in the universe, while point B, like the setting of the *Star Wars* epic, lies somewhere in a galaxy far, far, away. Now imagine that, from the perspective of A, a particle/antiparticle pair comes into existence precisely on the de Sitter horizon of the universe. Just like the particle/antiparticle pairs at the event horizon of any black hole, one goes over the horizon and in effect leaves the universe (leaves A's observable universe). The other becomes a widowed particle and flies away as a photon of de Sitter radiation. Billions of years later, it may even pass through point A.

Now, if we assume that point B lies more or less between point A and this same particle/antiparticle pair, what is the situation from *B's* perspective? The de Sitter horizon of B will not be the same as that of A. In other words, the observable universe from the perspective of B will not be the same as that from the perspective of A, although there will be a degree of overlap. But from B's perspective, the particle/antiparticle pair that, for A, sat exactly astride the de Sitter horizon is far closer than the horizon and neither member of the pair is in danger of crossing outside of the observable universe and leaving the other as a "widow". Instead, both come into being and annihilate almost immediately, each lasting for only the minutest fraction of a second. Following this line of thought, we arrive at the seemingly absurd conclusion that for A one of the particles becomes real and lasts for billions of years while for B both particles remain virtual and last for less than the blink of an eye!

A letter from the writer to Professor Leonard Susskind finally shed light on this puzzle. In his answer, Susskind first reiterated that, according to the uncertainty principle, the larger the energy of a particle pair the shorter its duration. For a high-energy pair astride the de Sitter horizon, the duration will be so short that only a microscopic distance will be covered, but the ones of interest with respect to the present problem are those with no rest mass and very small energy. These are the ones which escape as de Sitter radiation from the perspective of a point at the center of the sphere whose surface is constituted by the de Sitter horizon. They will indeed persist for billions of years from this (and only from this) perspective, but they will forever be undetectable—redshifted until their wavelengths are measured in billions of light years.

A slightly different, albeit closely related, way of looking at this might be to consider the time dilation effect as applied to A's widowed particle. The annihilation of this particle, located as it is right at the de Sitter horizon, takes essentially forever from A's perspective, even though it is all over in an instant from that of B where it remains far from the horizon and not located in a region of space that is expanding at the speed of light. The annihilation will likewise take but the briefest instant for the particle/antiparticle pair itself, because the pair sees itself, not astride the de Sitter horizon, but at the center of its own observable universe.

Whether raising time dilation here helps understanding or further confuses I am not sure, but in any case, enough has probably been said about this to reveal yet another instance of profound "weirdness" in the universe!

## Cosmic Strings

But there is even more.

As if black holes and cosmic horizons are not weird enough, most theories of the first few fleeting moments of the Big Bang predict the existence of exceedingly strange entities known as cosmic strings. These must not be confused with the strings of string theory, but are examples of topological defects in space itself, somewhat analogous to the cracks that form on the surface ice of a freezing pond. They are termed "line-like", which in plain language means one-dimensional for all practical purposes. A string is expected to have a diameter about one trillionth that of a proton, but coupled with a density such that just eight miles (10 km) of the same string weighs about as much as the Earth. And these strings, thanks to the expansion of space, stretch out across the visible universe and have in fact been suggested as the seeds for later aggregations of matter in the form of the observed lines of super clusters of galaxies. It is truly amazing to think that the largest structures in the universe—the great galaxy clusters and the lines or "walls" which these align to form and which appear like membranes of great cosmic cells throughout the universe—may have been shaped by such ultra-thin features as cosmic strings (Fig. 2.10).

Unlike black holes however, there is very little observational evidence that these things actually exist. That is not in itself too surprising, as they are predicted to be very rare. Still, they have been suggested as possible contributors to the missing mass of the universe. There is even a theory that some strings may have negative mass; a truly weird concept.

How could such a strange beast as a cosmic string be observed?

One method would be through gravitational lensing and, indeed, there was a suspicion that just such a lens may have been observed back in 2003 by a team led by M. Sazhin. The suspected lens consisted of two very similar galaxies located very close to

FIGURE 2.10 Galaxy cluster PKS 0745_665 imaged at X-ray, optical and radio wavelengths. X-ray image displayed as *purple*, optical as *yellow*. At center of cluster is a giant elliptical galaxy observable at radio wavelengths. This cluster is about 1.3 billion light years away (*Credit*: NASA Chandra)

each other in the sky. The suspicion was that, like the Einstein Cross for instance, these may have been two images of the same object, though in this instance produced by a very long and narrow lensing object (a string) rather than an intervening galaxy. Follow-up observations employing the Hubble Space Telescope in 2005 revealed, however, that the two galaxies really are individual objects and not two images of the same galaxy.

Gravitational lensing by strings should also produce duplicate images of fluctuations in the CMB, however no candidates were detected in a 2013 examination of data from Planck.

At present, the most promising evidence for the existence of cosmic strings is a curious observation of the so-called "double quasar" Q0957+561A, B made by a Harvard team led by Rudolph Schild between September 1994 and July 1995. The quasar itself was discovered in 1979 and is actually a gravitationally lensed double image of a single quasi-stellar object. The lensing agent is an intervening galaxy located about 4 billion light years from Earth and the lensing is such that one image of the quasar arrives some 417.1 days later than the other. This means that an intrinsic fluctuation in the light of the quasar, to which these objects are prone, will affect one image 417.1 days before it is seen in the other. But what the Schild study revealed was that during the period from September 1994 until July 1995, the time delay ceased. Any change in one image was immediately apparent in the other. Four events were recorded during this period, and none showed the expected time delay between images. Schild and his team believe that the only explanation for this is a cosmic string passing between Earth and quasar during that period of time. The string would need to be travelling at very high speed and oscillating with a period close to 100 days.

Although somewhat off topic, mention of the "double quasar" reminds us that in 1996, Schild and his Harvard team detected a fluctuation in the image of this object that appeared consistent with a microlensing event caused by a planet of three Earth masses within the intervening galaxy. Although this claim is controversial and unable to be confirmed, if it is correct, this would mean that a planet some *four billion light years* away has been detected! That is truly a planet in a galaxy far, far away (Fig. 2.11).

It is to be hoped that evidence for cosmic strings may come from another source in the near future. With the setting up of gravitational-wave observatories, it is possible that waves from oscillating strings might be detected. Whether these observatories will be sufficiently sensitive to detect waves from these sources remains, however, to be seen.

Is the Universe a Hologram?

Strange as it may seem, this is a question being seriously asked by physicists nowadays.

FIGURE 2.11 Double quasar Q0957+561 A, B (*Credit*: ESA/Hubble & NASA)

The path was prepared for serious consideration of this question by a way of interpreting the physical universe in terms, not of ultimate particles of matter, but as a system of information. This trend was inaugurated by J. Wheeler of Princeton University, but information theory itself, on which Wheeler's approach depends, originated back in 1948 as presented in papers by applied mathematician C. E. Shannon. It was Shannon who introduced that measure of information content known as entropy. The concept of entropy itself had long been employed in thermodynamics where it was popularly described as the measure of disorder in a physical system. A more precise account was given, however, in 1877 by Austrian physicist L. Boltzmann who characterized it in terms of the number of distinct microscopic states that particles composing a parcel of matter can be in without altering the external appearance of that parcel of matter. Shannon's discovery was his quantification of information content (in, say, a message) by employing a similar concept. The Shannon entropy of a message (for example) is the number of binary digits (*bits*) needed to encode that message. Both Boltzmann's thermodynamic entropy and Shannon entropy are conceptually equivalent. The number of arrangements counted by Boltzmann entropy reflects the amount of Shannon information required to bring about any specified arrangement of particles.

Now, according to thermodynamic theory, the entropy of an isolated physical system can never decrease. At most it might remain constant, but the more usual situation is for it to increase; something seen as the increasing disorder or dissipation of a system over time. This is the famous Second Law of Thermodynamics, sometimes stated to be the physical law having the greatest impact outside the confines of physics strictly so called. It is of the utmost importance to fields such as physical chemistry and engineering and appears to be correct beyond all reasonable doubt.

And yet black holes, it would appear, violate it. Or maybe not so much violate as transcend or simply ignore it. Whatever term we may use, the fact of the matter is that any parcel of matter falling into a black hole just disappears from the observable universe, taking its entropy with it! This point first appears to have been emphasized by Wheeler several decades ago.

Nevertheless, further research proved that the situation was not quite so straightforward. In 1970, Stephen Hawking and Demetrious Christodoulou (at that time a graduate student of Wheeler's) independently proved that in various situations—black hole mergers for instance—the total area of the event horizons never decreases. This in turn inspired Jacob Bekenstein of the Hebrew University of Jerusalem and himself a former PhD student of Wheeler's, to propose 2 years later that a black hole has entropy proportional to the area of its event horizon. When matter falls into a black hole therefore, the increase in black hole entropy always compensates or overcompensates for the apparently lost entropy of the disappearing parcel of matter. Bekenstein generalized this into the Generalized Second Law of Thermodynamics (GSL) viz. the sum of black hole entropies and the ordinary entropy outside the black holes cannot decrease. One immediate application of the GSL concerns Hawking radiation. Although this causes the total area of the event horizon and therefore its entropy to decrease, the GSL shows that this is more than compensated for by the increased entropy of the emergent radiation.

Further theoretical work by R. D. Sorkin of Syracuse University in 1986 determined that black hole entropy is precisely one quarter of the area of the event horizon as measured in units of *Planck areas*; one Planck area being the square of a Planck length. Now, the Planck length is not the length of a piece of wood, but is generally considered to be the smallest length for which the Heisenberg Uncertainty Principle allows physical meaning. It is named in honor of Max Planck, the father of quantum theory and has the incredibly tiny dimension of just $10^{-35}$ m.

Needless to say, one quarter of the area of the event horizon measured in terms of these tiny units amounts to a very large amount of entropy. Bekenstein calculates that even a hypothetical mini-black hole just one centimeter in diameter would encode approximately $10^{66}$ bits of information; equivalent to the Boltzmann entropy of a cube of water 10 *billion* kilometers wide!

Surprisingly, the above shows that information capacity depends on the surface area rather than on the volume of an object; something which agrees very well with theoretical research by Utrecht University's G. 't Hooft together with L. Susskind of Stanford University. This is where the notion of holograms enters the discussion. Holograms, as we know them in the everyday world, are photographs of a special kind in which all the information of a three-dimensional scene is encoded into a pattern of light and dark areas on the two-dimensional surface of a piece of film. When illuminated in a certain way, a full three-dimensional image is generated. What 't Hooft and Susskind demonstrated is that an analogue of this applies to a full physical description of any system occupying a three-dimensional region. In short, they propose that a physical theory defined only on the region's two-dimensional boundary will completely describe the three-dimensional system. The information content of the entire system does not exceed that of the description of the two-dimensional boundary. This all sounds very much like the conclusion of Sorkin, Bekenstein and Co. concerning the information content of black hole event horizons!

Now here is something truly mind-boggling. As the de Sitter horizon of the observable universe is also an event horizon, its entropy will also be equal to one quarter its area as measured in Planck areas. Now that is an enormous amount of information.

It is, indeed, all the information in the universe encoded on the furthest horizon. In other words, it would appear from this line of thought that the universe can be conceived of as a projection of a hologram encoded on the de Sitter horizon.

This sounds just too way out to be taken seriously, yet if it is true, certain observable consequences follow. As Fermilab particle physicist Crain Hogan noted, if the universe is a hologram, the quantum fuzziness of space should be detectable. This fuzziness arises from the prediction that no measurement smaller than the Planck length is physically meaningful, as we saw above. The Planck length is, of course, a very small distance, $10^{-35}$ m, which comes out at about 100 billion, billion times smaller than the diameter of a proton. At this level, space is widely thought to dissolve into a foamy, grainy form somewhat like tiny pixels. However the dimensions of these "pixels" are such as to place this "fuzziness" or "graininess" beyond any hope of direct observation. Nevertheless, if all the information in the universe is encoded on the de Sitter horizon, there cannot be more "pixels" within the three-dimensional space (or four-dimensional space/time to be more accurate) than is already encoded on the two-dimensional de Sitter horizon. Therefore, although the Planck lengths on the horizon are as small as $10^{-35}$ m, their projections into the universe of space and time will come out as large as $10^{-16}$ m. Now, that may just be large enough to detect. Indeed, for a while it looked as though they had been detected already!

This suggestion was raised in 2009 when Hogan began to consider the possibility of these space pixels showing up as interference in experiments being performed by the gravitational wave detector GEO600 near Hanover. Gravitational waves—yet another consequence of Relativity—may be thought of as ripples in the fabric of space and time caused by violent events such as the collapse of stars into black holes or collisions between extremely massive objects. They have yet to be detected directly, although the orbits of binary neutron stars have been observed as slowly decaying in just the way expected if orbital energy is being lost through the emission of these waves and, in March 2014, strong evidence for primordial gravitational waves was found in polarization measurements of the Cosmic Microwave Background. Still, scientists would dearly love to observe the waves directly, and it is here that detectors such as GOE600 are brought into operation. These are in essence extremely sensitive rulers. If a gravitational wave passes through the instrument, it will alternatively stretch space in one direction and squeeze it in another and this disturbance will hopefully be enough for detection by a laser beam fired through a half-silvered mirror known as a beam splitter. As its name suggests, this splits the laser beam into two components, which pass down two long perpendicular arms before bouncing back to merge again at the beam splitter. The returning beams create an interference pattern of light and dark regions that either reinforce or cancel each other. Any slight change in the relative lengths of the arms down which the light rays travelled would be betrayed by a shift in the positions of those regions.

Hogan worked out that, if the grains of the universe are as large as the holographic theory predicts, rather than being strictly Planck areas, the beams should experience a constant degree of buffeting which would show up as an unavoidable interference. After deriving the magnitude of this effect, he contacted the GEO600 team and was fascinated to find that the team had indeed been picking up a random jitter that looked exactly like the interference that Hogan predicted!

Unfortunately, these results do not appear to have been confirmed elsewhere and even some of the jitter at GEO600 has apparently disappeared. In fact observations of remote gamma-ray sources in recent years strongly imply that space is not grainy down to distances far *smaller* than the Planck length; a truly surprising result that may have great ramifications for a satisfactory theory of gravity. More will be said about this in a later chapter.

So the situation is still very much in flux at this time and all we can do is look forward to more work on this subject, both theoretical and experimental.

Yet, supposing that the theory does ultimately prove correct, what would that mean—really mean—for our understanding of the universe? It hardly needs saying of course, that it would be a major breakthrough in physics and cosmology and would undoubtedly lead on to new discoveries and new theories. It may even bring gravity into a unified field theory and get rid of the troublesome singularities that General Relativity predicts at the core of black holes. But what would it mean for our more workaday down-to-Earth understanding of the world around us?

Just saying the universe is a hologram seems to imply that it is somehow not real. But what can real and unreal even mean in this context? In our everyday thought, real means the opposite of unreal, imaginary and such like (and vice versa). My apparent perception that the image of the rising Moon is larger than that of the Moon when high in the sky is an illusion and therefore unreal. It can be shown by simple experiment that the images are actually equal in diameter. My apparent perception of an enlarged image does not correspond with reality, as we say. Likewise, if I am running a high fever and think that a spider is crawling across the floor toward me, I am said to be hallucinating if no such spider exists in reality. In both cases, although in differing degrees, what I seem to experience does not correspond to anything actually existing out there. Similarly, if the three-dimensional statue that I see before me is a holographic image, once again I appear to be experiencing something that is not out there; not, at least, in the form that I believe I am experiencing it.

All of these instances of illusion, hallucination or miss-perception in general are judged to be unreal by comparing them with a single standard for reality—the external world, which ultimately encompasses the universe as a whole. But by what standard can we compare the reality of this very universe; this ultimate frame of reference which itself acts as our final standard of physical existence?

This is not to say that we are compelled to be naive realists, i.e. committed to the epistemological doctrine that we always accurately perceive the surrounding world, but it does mean that our overall experience of the world inclines us to believe in an objective reality. Only by belief in this can the concept of illusion have any meaning. To say that all is illusion is another way of saying that nothing is illusion; that the spiders of a fever patient or the writhing snakes seen by someone suffering the delirium tremens are somehow as real—or, equally, as unreal—as the trees and the stars!

If the universe is a hologram, it is also physically real. Perhaps it is this line of thinking that has caused some scientists to opine that even if the theory is correct in the sense that the universe may conceptually and mathematically be expressible in the same way as a hologram is expressible, that still does not mean that the universe *is* a hologram. All that can be said is that the universe is like a hologram. At least, when considered in this way, it is like a hologram. It may be treated theoretically as such, but that does not necessarily mean that it literally *is* a hologram.

The philosophical considerations here are parallel to the statement that we read at times in elementary accounts of atomic theory, namely, that solid objects are really like swarms of gnats. We of course appreciate the point being made—that analyzed in atomic terms a piece of ordinary solid matter consists of very small particles and lots of empty space—but that does not alter the fact that as experienced in our familiar world, pieces of solid matter are nothing like swarms of gnats. The real problem with these statements comes when it is said that a person walking across a plank of wood is really like someone walking on a swarm of gnats. Somehow the plank of wood gets analyzed in atomic terms, but the person and the act of walking remain in familiar-world language. A moment's reflection reveals just how absurd this statement really is, but it (or something very similar) has been read and heard so often that it is scarcely given any thought!

With these considerations, we leave the holographic account of the universe. Whatever the final word on this theory, we can be sure that its discussion will continue to raise issues of both a physical and a metaphysical nature that, even if the theory turns out to be incorrect, will further deepen our understanding of this amazing universe in which we live.

# 3. The Shape, Size, Age and Origin of the Universe

## Has the Universe a Shape and a Size?

Back in Chap. 1, we noted that the universe could have three basic geometries, Euclidean, Riemannian or Lobachevskian. The first is the familiar plane geometry as taught in school; the second is the geometry of a closed hypersphere (like a sphere, but existing in four rather than in three dimensions) and the third a pseudosphere (sometimes likened to a saddle, although once again the extra dimension makes this a not-especially-accurate way of picturing it). Triangles in the first type of space have angles that sum to 180° just as we were taught in elementary school, but the sum of all the angles of triangles in the second and third types of space are either greater than 180° (for Riemannian) or less than 180 for Lobachevskian geometries.

Frequently when these possibilities are introduced in elementary books on astronomy, it is stated that the first and last possibility each implies that space (in other words, the universe itself) is of infinite extent. The second, being closed, implies a finite universe. One could travel literally forever through a Euclidean or Lobachevskian universe and never return to one's point of departure. On the contrary, the home port would just get further and further away. On the other hand, if one set out on a cosmic journey in a Riemannian universe, eventually one would return to the journey's starting point. It has even been said that if one had a sufficiently powerful telescope, one would ultimately see the back of one's own head. Actually, it doesn't quite work that way thanks to the finite speed of light. What one would see is a point in space that will come to be occupied by the back of our head billions of years after the light now being observed left that point. This is of no concern for us at present however.

Nevertheless, during the course of the last several decades, this simple picture has been complicated somewhat by considerations of the overall shape or topology of the universe. The fact of the matter is, while Riemannian space must be finite, Euclidean and Lobachevskian spaces may be either finite or infinite, depending upon how the universe is bent and twisted. Take a flat space for instance. At first thought, this would appear to imply an endless plane stretching away forever in all directions, just as imagined in the introductory paragraphs above. However, what if the universe is (to take one of many possibilities) shaped like an English football having a number of flat tiles joined together into a polyhedron? Each "tile" or face is still flat—a triangle drawn on any will have 180° as the sum of its angles, just as surely as one drawn in a Euclidean space of infinite extension. But a football-shaped universe will be finite. Similarly, if the universe is shaped like a horn or bugle, the geometry of space will be Lobachevskian but the extent of the universe, finite.

Several decades ago, a department store in Sydney, Australia, erected an interesting display alongside the escalator moving between the first two floors. As one travelled up the escalator, a mirror wall on one side gave the effect of a row of escalators trailing off into an indefinite distance. People riding the escalator could glance to one side and see what looked like an infinite set of escalators—and images of themselves riding these—receding into an indefinitely far distance. This hall of mirrors effect is well known, but during the latter years of last century, a few cosmologists thinking outside the proverbial box raised the possibility that our universe might have more in common with a hall of mirrors than we had hitherto realized! What if the seemingly endless array of galaxies stretching outward, apparently forever, was simply an illusion like the seemingly endless array of escalators in the Sydney department store? Maybe the actual universe is quite small—possibly smaller even than the apparent sphere of distant galaxies which we observe—and that what appears to us as a large, or even infinite, universe is simply an illusion; images of the real universe repeated over and over again!? Unlike the images of the escalator and its riders however, the distances involved and the finite velocity of light would mean that each image of the universe would date back to a different earlier time, so effectively the cosmic hall of mirrors would also be a sort of time machine showing the universe at increasingly earlier epochs. Referring back to the mirrored images of the escalator and the observer riding the same, this is analogous to the reflected images showing earlier and earlier images of the same observer who would then, of course, be progressively younger as he looked down the line of reflections. The "nearest" images would show (say) a middle-aged person, slightly more distant ones, a young adult, then a teenager, a child, a baby and finally an embryo; the latter corresponding, in the universal example, to the Cosmic Microwave Background. In the real universe, this process of cosmic evolution means that an astronomer could not simply look for an immediately recognizable image of (what we might loosely term) the "local" universe at more distant/earlier epochs. Nevertheless, in a small universe, patterns in the CMB should indeed be discernible and have been eagerly sought, especially since the launch of the Wilkinson Microwave Anisotropy Probe (WMAP) in 2001. No such patterns have been detected however, indicating that the real universe is probably not smaller than the observable universe and that we are not seeing a cosmic hall of mirrors when we gaze out into distant space and deep time. Nevertheless, that does not mean that the universe is not finite or that topology does not need to be taken seriously. There was never any strong reason for thinking that a finite universe would be small enough to yield up its true dimensions in this way, although that possibility was great enough for the idea to be observationally tested.

Even though the WMAP results apparently rule out a very small universe, the map of the CMB obtained by this satellite (and, indeed, that of its predecessor COBE in 1991 and, according to results available at the time of writing, its successor Planck launched in 2009) may not entirely agree with an infinite universe either.

All three probes revealed the CMB as mottled with warmer and cooler regions, reflecting variations in the density of the very youthful universe. This much was expected and, indeed, in practically every respect, the pattern fits what the inflationary Big Bang model of the universe predicted. Yet, the agreement with prediction is not perfect. Although the expected small-scale fluctuations match predictions very well, the largest fluctuations do not.

Fluctuations that subtend angles of 60° or larger across the sky are a lot weaker than expected.

What does this imply? No definitive and uncontested answer has as yet been given, but a number of scientists quickly drew attention to the discrepancy between this lack of large-scale fluctuations and what was expected for an infinite universe. What was observed provided a far better fit to a finite universe, although alternate explanations cannot be ruled out (for instance, it has been proposed that what ended up as the larger fluctuations were generated earlier in the inflationary era and subsequently became "ironed out" by the inflationary process itself).

In a sense, the universe can be likened to a ringing bell and the observed temperature fluctuations as the notes produced by this bell. We know that small bells produce high-pitched notes whereas large bells toll deeper tones. Simply stated, an old-fashioned dinner bell is not large enough to achieve the large-amplitude vibrations required for the deep tone of London's Big Ben! The same is true for the universe. If the universe is relatively small (albeit not necessarily so small as to fit inside the confines of our sphere of observation) the largest fluctuations, like the deep tones of a large bell, cannot be produced. Analysis of the WMAP data concluded that there is just one chance in 3,000 that large-scale fluctuations as weak as those indicated could be produced in an infinite Euclidean universe.

Nevertheless, even if it could be shown that these results did imply a finite universe, as they stand they say nothing about its specific topology.

One early model of a possible finite universe put forward by University of California astrophysicist Joseph Silk and Cambridge theoretical physicist Janna Levin was that of a dodecahedral universe, like the English football mentioned above. According to their scenario, any object that travels away from its point of departure in a straight line will eventually return from the other side of the universe, having been rotated through an angle of 36° in the process.

An even more bizarre model was proposed by Frank Steiner of the University of Ulm. Steiner suggested that the universe may possess what is known as a Picard topology, that is to say, that it may be hyperbolic and yet finite and shaped like a horn or trumpet or, we could also say, somewhat like the flower of an arum lily. The universe that we know occupies the flared end of the trumpet or the petals of the lily, which ever image we might prefer. But where things get really weird is in the region away from the flared end. As we travel further and further down the throat of the horn of stem of the lily, space becomes increasingly folded until eventually it stretches out into a line of infinite length.

Although neither living beings nor even places such as planets suitable for the habitation of such beings could exist too far into the throat of a Picard universe, if we suspend disbelief for a moment and try to imagine what life down the Picard throat might experience, we come up with some truly bizarre possibilities. For instance, imagine two children sitting at desks, one at the front and the other at the back of a classroom. The teacher asks them to draw a triangle, measure each of its angles and then add the results together. Each child dutifully performs the task and each comes up with a different answer, neither of which is 180°! Neither child is incorrect, but if we imagine that their classroom is very far down the throat of the Picard horn, the curvature of space will be more extreme at one end of the room than at the other. A little further still down the throat, and a person looking straight in front would see the back of his own head. The space in that region would be so extremely curved as to cause a beam of light to trace a full circle.

Bizarre though it may seem, the pattern of the CMB is not inconsistent with the real universe having a Picard topology. Indeed, supporters of this model drew attention to an apparent elliptical shape of the smallest spots in the microwave map, arguing that such a shape was more consistent with a Picard topology and hyperbolic space than with flat space. Flat space should produce round spots, although distinguishing between the two shapes is difficult in practice and the slight departure from circularity noted is less than overwhelming evidence that this bizarre topology describes the real universe.

In theory, a double Picard horn is also possible. In this model, the line into which the horn contracts is of finite length, eventually flaring out into another horn and, presumably, another universe. Such a shape would leave a distinct pattern in the CMB however and this does not appear to be present.

Another possibility is for a number of separate Picard horns, maybe having the straight-line regions converging at some point in hyperspace like the stems of a bunch of flowers tied together in a bouquet! Such a scenario is, however, pure speculation not based on any evidence.

By the way, it is sometimes stated that the Picard topology was named for the starship captain of that name in *Star Trek, the Next Generation*. Sorry Trekkies, but that is not correct! It was actually named for the French mathematician Charles Emile Picard (1856–1941).

## Donuts, Axes of Evil and Other Anomalies!

The CMB has, in general, provided much confirming evidence for the Big Bang cosmology. Yet, as already stated, agreement is less than perfect in some of the details where, as the old saying goes, the Devil lies! As we saw, the largest features are a lot weaker than expected and it is this, plus (let's be frank) the philosophical difficulty which many of us have with infinity, that led some cosmologists to take seriously the possibility of a finite universe. Others suggested that the apparent discrepancies might be simple artifacts and hoped that they would go away as further data, from more sensitive probes, were gleaned.

That hope proved forlorn. The WMAP probe actually strengthened the apparent findings of COBE and (although still a work in progress at the time of writing) results thus far released by the Planck team indicate that this latest probe—the most sensitive to date—will further strengthen the earlier results (Fig. 3.1).

In addition to the comparative lack of large patterns, increasingly refined analysis of the chart of the CMB has come up with a very unexpected feature. The all-sky image revealed that one direction in space, one hemisphere of the microwave sky, appears lumpier than the other. Not only that, but when the finer variations in the microwave background were taken out of the picture, the remaining large-scale ones (known as the quadrupole and octopole) formed an alignment across the sky. This was most unexpected; so much so that the feature was dubbed the axis of evil, borrowing the term from George W. Bush's characterization of an alleged alliance of nations supporting terrorist attacks on western

FIGURE 3.1 The Planck spacecraft (*Credit*: NASA)

democracies. Needless to say, the cosmological axis is not really evil and comparison with terrorism was a little harsh, but the use of this term, even in jest, shows the extent to which its discovery unnerved the scientists who found it! (Fig. 3.2)

On the face of it, this pattern in what is really the glow of the universe's ultimate horizon—the glow of cosmic dawn if we wish to wax poetical—throws a very large spanner in the philosophical works underlying much scientific thinking. We have already referred to the Cosmological Principle which, as recalled from Chap. 1, states that the universe, on a large scale, appears the same in all directions. This principle is accepted as an act of faith, even though there are some grounds for being critical of it. As we saw in the first chapter of this book, it certainly does not work on small or moderate-sized scales and breaks down when the fourth dimension of time is added i.e. when it becomes the Perfect Cosmological Principle which, as we saw, predicts the Steady State model of the universe and clashes with any idea of cosmic evolution as

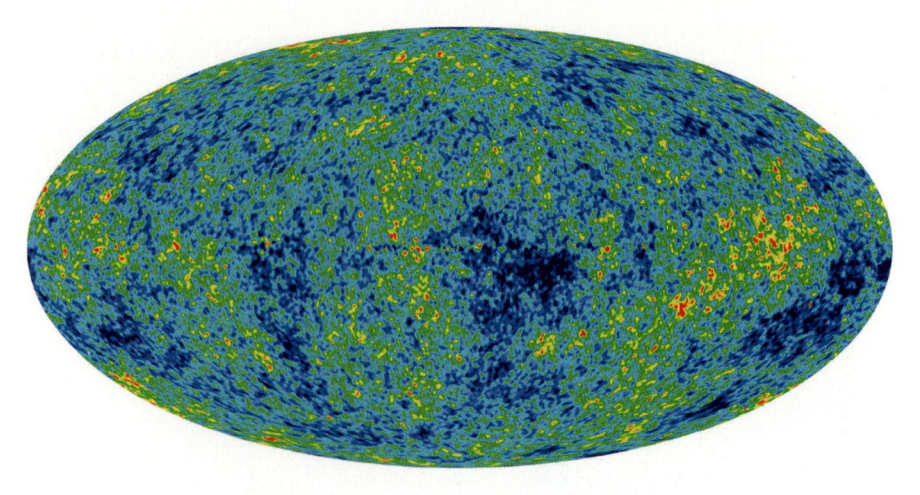

FIGURE 3.2 Whole-sky image of the Cosmic Microwave background (CMB) (*Credit*: NASA)

necessitated by the Big Bang. If it should eventuate that the CMB violates the CP, the latter would need to be abandoned altogether; something that cosmologists would be very loathe in doing.

The so called axis of evil remains a mystery at the time of writing. Hopefully, further analysis of Planck data will wring out an answer. It would be nice to think that the problem may even have been solved by the time you read these words, although I must admit that I am not very optimistic about this. In any case, the issue is such an important one that much analysis and theorizing will be inspired by it, as indeed, it has been already.

While an answer on which most agree lies in the future, several interesting proposals have already been made by various scientists and, for all we know, one of them may even prove correct.

At the conservative end of the spectrum, it is argued that the apparent anomalies in the CMB are just that—apparent. The chance of the observed alignment being pure coincidence has been estimated as around one in one thousand. Although this is small, it cannot be doubted that rare co-incidences do happen. We are reminded of an incident during the London blitz of the Second World War when a bomb fell through the roof of a house and failed to explode. A little later, a second bomb fell through the hole made by the first … and also failed to explode. What are the chances of this happening? Less than one in a thousand I would guess.

## The Shape, Size, Age and Origin of the Universe 121

A somewhat less conservative position—albeit one that does not call into question firmly-held philosophical presuppositions—is that the axis is real enough in the sense that it has a physical cause and is not simply a statistical fluke, but that its origin lies not within the CMB itself but in the intervening space through which the ancient microwaves travel. The "axis" is, in short, an example of gravitational lensing, like the ones about which we spoke in the previous chapter, albeit on a far grander scale. Indeed, if this is the explanation, it involves the largest example of the phenomenon in the observable universe; the lensing of the CMB itself by means of a massive cluster of galaxies. The prime suspect for this cosmic prankster is the so-called *Shapley Supercluster*, more prosaically known in the official catalogues as *SCI 124*. This is a humungous system located some 650 million light years away in the constellation of Centaurus and consisting of approximately 100,000 individual galaxies. Such an enormous concentration of mass is quite capable of performing this trick and several astronomers suspect that this might indeed be the solution to the axis of evil mystery. Many, I am sure, want to believe that it is! By the way, it is fitting that this super-giant cluster of galaxies should be found in Centaurus, as this constellation contains a number of other very interesting objects as well. The nearest star to our solar system (the Alpha Centaurus triple system) lies in that constellation, as does one of the largest and brightest of our galaxy's globular star clusters (Omega Centauri) as well as the powerful radio galaxy NGC 5128. The Shapley Supercluster has some interesting company in our sky although, of course, it lies far beyond these other objects.

At the other end of the spectrum of speculation (dare we call it the left wing?) we have theories involving a radical change in our ideas about the universe and maybe even the introduction of new physics. It is at this point where suggestions of a finite universe raise their head once more. Max Tegmark, for example, found that the apparent anomalies in the CMB could be explained if the universe is toroidal in shape, somewhat like a donut. This may sound like something from Homer Simpson, but it is not ruled out by the equations of General Relativity. It has even been suggested that the CMB patterns can be explained, not just by a donut universe, but by a donut that is also spinning.

If the CMB anomalies really do require such a radical explanation, problems will be raised for inflationary theory. As we have seen, inflationary theory is not without its difficulties, but it also nicely accounts for much that is hard to explain on any known alternative and cosmologists are not likely to abandon it without strong evidence to the contrary. If inflation is at least basically correct, one consequence is that our observable universe is just a minute spot within a vastly larger cosmos. Not all supporters of inflationary theory insist that the "total" universe is necessarily infinite in the full sense of the term, but it may as well be for all practical purposes. The point is that if the actual universe turns out to be not just finite, but small enough for its finitude to be manifested in the CMB, inflationary theory, as we now understand it, will be in need of some modification. That is something that most scientists would be reticent about believing unless absolutely forced to by reason of very strong evidence. At present, no such overpowering reason has been found. The donut and other finite topologies have some evidence that can be construed as being in their favor, but it is equally possible that this evidence can be interpreted in other ways. Indeed, this is what the more conservative explanations for the "axis of evil" and the apparent difference between the hemispheres of the microwave sky are hoping to accomplish.

The difficulties raised by a literally infinite universe will be taken up shortly, but at this point, mention should also be made of another anomaly detected in the CMB map that adds to the difference between hemispheres and may or may not be relevant to the "axis of evil". This is an apparent cold spot in the direction of the southern constellation Eridanus. Slightly warmer and slightly cooler spots cover the microwave sky, but this one appeared a little too large and a little too cool to be overlooked. It was initially detected in WMAP data and at first thought (hoped?) to simply be an artifact of that probe which would disappear in the higher-resolution Planck data. In fact, it emerged clearer than ever in the Planck results! (Fig. 3.3)

Once again, explanations range from the conservative (simply a chance fluctuation), to the not-quite-so-conservative (maybe it marks the position of an intervening intergalactic cloud of dust and/or gas that is absorbing the radiation from the CMB

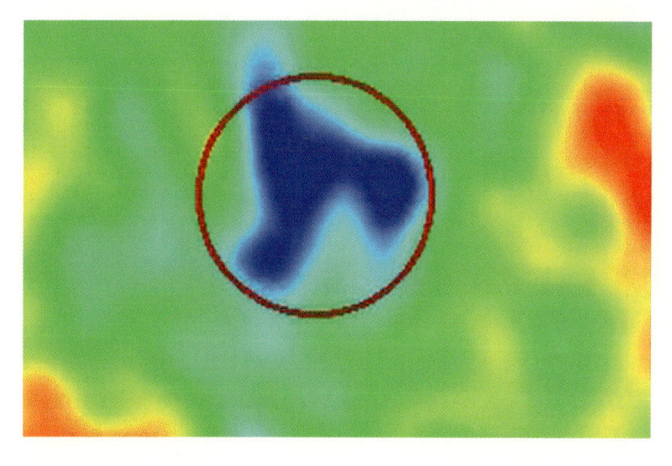

FIGURE 3.3 The enigmatic "cold spot" in the CMB (*Credit*: Rudnick/NRAO/AUI/NSF, NASA)

background) to radical proposals (for instance, the cold spot is a kind of giant knot in space known as a topological defect. Such creatures are predicted in some theoretical models). One suggestion from right out of the left field is that of Laura Mersini-Houghton of the University of North Carolina who has advanced the thesis that the cold spot is actually the imprint left by a universe, other than our own, which formed alongside ours. The influence of this neighbor universe is also, she suggests, responsible for the existence of the "axis of evil". More will be said about this in the following chapter, as the full explanation involves input from the controversial string theory, which we will look at in that chapter. Two remarks are pertinent at this stage however; one is the apparent correspondence on the celestial sphere between the CMB cold spot and an enormous void, some 900 million light years across, revealed by the Sloan Digital Sky Survey. This void lies about 8 billion light years from Earth and contains up to 45 % fewer galaxies than one would expect for a similarly-sized region of the universe having average galaxy density. The size and distance of this void is about what Mersini-Houghton predicted for a hole left by a separating universe. She also predicted that a similar cold spot and void should be present in the opposite hemisphere. The status of this feature is uncertain at the present time. Its detection has been claimed as the result of some analyses of the CMB, although other scientists have cast doubt on the accuracy of the methods used in support of this claim. For the present therefore, the explanation of the cold spot as evidence of another universe remains a very intriguing idea, but one that clearly needs more evidence.

## Infinite, Finite … or Both!

Can we imagine an infinite universe? It is certainly not easy. We can think of a universe so large that for all practical purposes, it is never-ending. Yet the gulf between this and a universe that really and literally goes on forever is one that our finite minds find impossible to bridge. Even if we believe it at the intellectual level, we still cannot truly appreciate it.

Yet, even if the universe at large is infinite, the universe as we observe it is clearly finite. As we look out into space and back into time, we see a younger and younger cosmos until, eventually we arrive at the wall of the CMB. Because the universe is finite in age, even if it should be infinite in extent, what we observe as our universe has boundaries. Indeed, even in a universe of infinite age (such as the Steady State Theory proposed) as long as space is expanding, we still cannot see out to infinity. We would continue to reach a boundary where the apparent recession of galaxies equals the speed of light. Beyond this, nothing can be seen.

The situation would be different in a universe that was infinite in extent, eternal in time, non-expanding and without a cosmological redshift resulting from any other cause such as the weakening of light with time. In a universe such as that, wherever we looked in the sky, there would be a galaxy located there at some distance from us. The galaxy may be near or it may be very far away, but somewhere in the direction of our gaze, a galaxy would be located. From this it follows that no truly dark region of sky could exist—photons of light would arrive from each and every point on the celestial sphere. Not even sections of sky where remote objects are blocked by relatively nearby clouds of cosmic dust would be exempt. These clouds would themselves be exposed to the universal flux of photons which are then absorbed by their constituent particles and, as energy cannot be destroyed, eventually re-emitted as these are warmed by the absorbed energy.

The upshot of this is that the entire sky would be a blaze of light. More than that, in a universe that is truly infinite in both space and time, the photon flux itself must be infinite. A static, eternal, non-evolving and infinite universe would therefore be a soup of photons, incapable of supporting any form of life!

This argument is commonly known as Olbers' Paradox (it appeared as a "paradox" when the universe was really believed to be eternal, infinite and unchanging), although the issue had been raised long before Olbers' time. It dates back at least to Edmond Halley, the famous colleague of Newton's who also acted as his patron in the publication of the Principia.

Fortunately, this is not the kind of universe that really exists!

Incidentally, we might think that because the age of the actual Universe is just short of 14 billion years, 14 billion light years should mark the distance to the boundary of the observable universe or (expressing the same thing a little differently) that the radius of the observable universe should be 14 billion light years. In fact, the radius is closer to 46 billion light years! But how can that be possible? We can't look back before the creation!

The solution lies in that weird fact spoken about earlier; the stretching of space. There is more than 14 billion light years of space to the edge of the observable universe because—like a toy balloon that has been inflated—the fabric of space itself has stretched since the Big Bang. In short, the light from remote galaxies now has more space through which to travel than it did when first emitted!

This observable universe—"our" universe—is sometimes referred to as the *O-sphere* or observational sphere. An infinite universe may therefore be looked upon as an infinite set of overlapping O-spheres. Now, this is where things start getting strange. Because the volume of each O-sphere is limited (albeit very large) and because the number of particles within this volume must, of necessity, be finite (albeit once again, very large) the number of possible combinations of these particles must likewise be finite (we may add ditto if we like). According to Max Tegmark, even if our O-sphere was filled with particles, these could only be arranged in 2 to the power of $10^{118}$ different combinations. Therefore, even if space really is infinite in extent, even if the number of O-spheres is truly infinite, there will only be a finite number of differences that can exist between them; that is to say, a finite number of possible combinations of their constituent particles. The weird conclusion is that, in a truly infinite universe, each O-sphere will be infinitely reproduced. There will be an infinite number of O-spheres exactly like the one in which we live. This means that there will be an infinite number of Earths exactly like ours, even peopled by an infinite number of people exactly like us, thinking exactly the same thoughts and doing exactly the same things as we do. But if that is not weird enough, there will also be an infinite number of not-quite-exact copies of our O-sphere. There will be infinite O-spheres exactly the same as ours except that Mr. Smith wears a hat today whereas in ours he wears a cap. There will be infinite O-spheres exactly like this one, except that this writer's hair turned grey with age instead of falling out. There will be an infinite number exactly like ours except that the rock which today falls off a cliff on the fourth planet from the millionth star in the outer arm of the Andromeda Galaxy manages not to topple until tomorrow. There will even be an infinite number of O-spheres exactly like this one except that two hydrogen atoms deep in intergalactic space and separated by millions of light years approached each other one millimeter closer ten thousand years ago than their counterparts in our O-sphere. And so on and on. Needless to say of course, these O-sphere clones will be separated by vast distances. According to Tegmark, an O-sphere containing your exact double will be located some 10 to the power of $10^{29}$ m away and an exact clone of our entire O-sphere will be 10 to the power of $10^{118}$ m distant.

If this is true, a strange paradox emerges which is best exemplified by a little science fiction tale. Deliberately overlooking the inconvenient restriction imposed by the absolute speed of light, let's imagine a Captain Kirk of the remote future navigating a starship which can not only zip at super warp speed around the galaxy, but even flit through entire O-spheres with the ease that a Saturday night joy rider speeds through neighboring suburbs. Captain Kirk leaves Earth orbit and heads out where no man has ever gone before. Picking up speed, he leaves our galaxy, then our local group of galaxies, then our O-sphere far behind. He traverses O-sphere after O-sphere, looking for the nearest that is an exact clone of our own. After a certain time has elapsed, he finds it (don't ask how, this is after all only a fictional story!) and with a little more searching, hones down on the Milky Way look-alike and finally on Second Earth. He arrives to a hero's welcome! His friends and family rush over to him, welcoming him back home.

"But" he stutters, "I have not been here before. I left another Earth in a distant O-sphere and have traveled across the vast cosmos to this exact duplicate of my home planet. The man you think you are welcoming is not the same as the one who left. I am an exact counterpart of that man!"

"Well" someone says, "let's test this. Do you remember me from college?"

"Well, you do look a lot like Joe Black."

"That Jim, is because I *am* Joe Black! Remember the time we played a trick on the physics tutor?"

"Oh yes, we put an anti-gravity device in his pocket. Took him an hour to come down off the ceiling."

"That's right Jim. It happened just like that."

And so on. Every memory Captain Kirk has of his life is confirmed by his friends on this "second" Earth. You see, as "our" Captain Kirk left "our" Earth, "their" Captain Kirk left "second" Earth and would now be experiencing exactly the same conversations after landing on "our" Earth as "our" Captain Kirk is having on "second" Earth!

Confused?

Well, try this thought. In what way can "our" Captain Kirk and "their" Captain Kirk be differentiated?

We might suppose that it would be similar to differentiating between two identical twins. But there is a good deal more to it than that. But to better appreciate the difference, let us get a little philosophical.

Philosophers speak of strange things called universals. These can be thought of as general (as distinct from particular) things, but we will not worry in this book about the numerous theories as to what they really are. After all, philosophers have been arguing about this since the time of Plato, so anything that we could say here is unlikely to be taken as being definitive! All we need to know is that philosophers recognize two types of universal; qualitative and relational. The first includes such things as colors (redness, for instance, is a qualitative universal) and the other intrinsic properties that a particular thing possesses. Relational universals, as the term implies, define a particular thing's position with respect to other particulars and, in the final analysis, to the remote masses of the universe.

But now back to our identical twins; let's call them Tweedledum and Tweedledee, with apologies for the lack of originality. We suppose that Tweedledum and Tweedledee are absolutely indistinguishable in all of their physical and even mental features. They share, in other words, exactly the same qualitative universals.

Suppose that the twins are rather elderly and each receives an age pension from the government. Suppose, further, that their uncanny likeness begins to arouse suspicion from the authorities as to whether the Tweedles are carrying on a scam. Perhaps there is only one Tweedle and he is drawing two age pensions—how are the authorities going to find out?

We imagine that (somehow, we need not bother ourselves with details) an investigator manages to plant a tracking device on Tweedledee and another, at a different time, on Tweedledum. The two devices are activated and each reveals a different location. Device A reveals one twin to be across the road when the devices are activated whereas device B shows its wearer to be over a mile away in the local park. Clearly, the two twins are different people, even though each can be adequately described qualitatively in terms of the same set of qualitative universals. What distinguishes them is a different set of relational universals. If, on the other hand, the two devices had always revealed the same location, the suspicion of dirty work would have been verified. Two people—or "two" anything—that can be completely described by both the same qualitative *and* relational universals are not two objects at all. They are a single entity.

Relational universals require a reference frame which, in the final reckoning, is the frame of the remote objects of the universe. Yet, in the case of two absolutely identical O-spheres, part of this identity *includes* the remote masses of the universe. Whereas Tweedledee and Tweedledum were identified by differing relationships within a broader frame of reference (which itself was part of the ultimate frame of reference provided by the remote masses), this cannot be so for identical O-spheres, because these do not lie within any encompassing frame of reference: Each O-Sphere *is* the encompassing frame of reference for every lesser frame of reference within it. Not only is Captain Kirk duplicated (actually, replicated infinitely), but the entire frame of reference within which he exists is likewise replicated. According to every conceivable observational test, they are identical. The philosophical issue raised here is known as the identity of indiscernibles. In other words, if A and B are indistinguishable in all respects—including relational ones—then can they be spoken of as *two* things at all?

Does it therefore make sense to speak of *two*—or an infinite number of—identical O-spheres? As far as Captain Kirk is concerned, his experience could just as easily be interpreted as his having circumnavigated a spherical finite universe and returned his starting point. These two alternative interpretations equally explain his observations; in other words, they are indistinguishable. So are they each, in a sense, correct?

This may be even more apparent if the universe is looked upon from the perspective of information content. From this point of view, the basic units of the universe are not a subatomic particles; they are information bits. The O-sphere is, essentially, a great store of information. Therefore, it could be said that the O-sphere which Captain Kirk leaves and the one in which he arrives are identical because the store of information is identical in each. Our intrepid space captain is simply re-encountering the exact store of information that he previously left behind!

We shall see in due course how some interpretations of quantum theory propose the existence of many worlds or alternate universes, not scattered through an infinite space, but as alternate spaces existing alongside our own. Interpretations of quantum physics are a little like alternate universes themselves, and we stress that the so-called "many worlds" interpretation is just one of the list, but the only point in bringing this up here is to point out that most proponents appear to interpret it not so much as an infinite assemblage of *other* universes as an assemblage of other versions of *this* universe. It is a rather strange psychological fact that where alternate universes are proposed in quantum theory, they tend to be presented in this light, while alternate O-spheres are nearly always expressed as truly "other" more or less similar clones of the known universe.

A similar issue arises is we imagine an oscillating universe or one that goes through an infinity of incarnations. In this case, our thought experiment has the intrepid Captain Kirk travelling not through space but through time, somehow passing unharmed and unhindered through the big bangs by which these universe cycles were born and the big crunches though which they died and were resurrected in new big bangs. Finally, just as before, he arrives in a universe—a cycle this time, not "another" O-sphere—that is exactly like the one he left. Exactly like or exactly the same? Surely, the latter, for the same reasons discussed above.

Again, we might think of the idea put forward by nineteenth century philosopher F. Nietzsche that the universe and all of its denizens repeat infinitely. In common with many scientists and philosophical thinkers of his day, Nietzsche imagined the universe to be infinite and eternal and ultimately composed of indestructible material atoms in perpetual Newtonian motion. But because these atoms could only arrange in a finite (albeit enormously large) set of configurations, he argued that each configuration must repeat infinitely, given an infinite extent of space and an eternal duration of time. Nietzsche saw this not so much as infinite replication of identical configurations as infinite repetition of the same configuration. This hypothesized eternal recurrence was presented as a kind of immortality. According to this philosopher, many, many millions of years hence, I will once again be writing these words and you will once again be reading them.

This interpretation differs from the version presented here in so far as it retains the same spatio-temporal framework. That is to say, it implies that in an infinite number of elsewheres within the infinite reaches of space I also exist at this present time and/or, at an infinite number of "elsewhens", both past and future but still within the infinite space of this universe, I have existed/will exist. Of course, the concept of O-spheres was a foreign one in Nietzsche's day as the discovery of cosmic expansion still lay in the future.

What are we to say about all of this? Does this expose another weirdness of the universe or, just maybe, are these conceptual difficulties and seeming paradoxes pointing us toward a different conclusion. Rather than being pointers toward a weird feature of the universe, are they instead presenting a reductio ad absurdum of the very concept upon which this weirdness rests, namely, that of infinity itself?

# The Shape, Size, Age and Origin of the Universe 131

## Is Infinity a Myth?

Infinity is at best a slippery subject, and many scientists who dare to cross the threshold into philosophy are not entirely happy with it. We could say that any concept that raises the sort of discussion as the above has got to have something wrong with it, and that is in effect what many philosophically inclined scientists are indeed saying. Thus, Janna Levin raised the point that if a mathematician finds that an equation gives an answer of infinity, she immediately concludes that something is amiss with the math. We all probably recall those arithmetic lessons at school when we ended up with a vulgar fraction whose denominator was zero. What did the teacher tell us to do? "Try again and get it right this time!" or something to that effect!

Why then, Levin asks, when the equation of the universe gives an answer of infinity according to certain models, is this accepted at face value and not as a warning that something, somewhere, has gone wrong with the theory? Does this, perhaps, tell us more about human psychology than it does about the universe; the psychology that says that the universe is so weird that infinity is the best answer after all?

Then there are the paradoxes raised by the amazing hotel invented in a thought experiment by mathematician David Hilbert. Imagine a hotel with an infinite number of rooms all filled with guests. In one version of the paradox, a new guest arrives. Can she be accommodated? It would appear that she can be, even though an infinite number of rooms have already been filled by an infinite number of guests. Any addition to infinity still gives infinity! In another and more extreme version of the paradox, an infinite number of guests arrive at the hotel. But they are also immediately found accommodation. Moreover, despite the infinite guest intake, the hotel still has an infinite number of vacancies and can readily accommodate another infinite influx of new guests!

If that is not paradoxical enough, suppose now that an infinite number of the guests decide to leave the hotel and go elsewhere. How many rooms are now left vacant? The answer is ... an infinite number: Exactly the number left vacant before the infinite number of guests left. And how many guests is the hotel left with after an infinite number leave? An infinite number of course! The numbers haven't changed.

One is tempted to deny that infinity is really a number at all, despite mathematical arguments to the contrary. This was the position taken by philosopher John Anderson. As Anderson saw it, declaring the universe to be infinite is really just a way of saying that the universe never ends; that one can never come to the end of it, simply because there is no "end of it" to reach! Calling infinity a "number" implies that it has a set value, which it clearly does not have; not, at least, in the usual sense. Perhaps infinity can be compared to an asymptote that is approached but never reached. But that is not satisfactory either as, unlike an asymptote, we never actually come any closer to it.

At one level Anderson's position does appear to avoid the paradox of Hilbert's Hotel. If infinity is not a number with a set value, there is no set number of hotel guests to which a further number is added. But is that really the end of the story? Read through the above several paragraphs again but replace "an infinity of ..." for each occurrence of "an infinite number of ..." (as Anderson would suggest that we do) and see if you think that the Paradox is made less paradoxical.

If what we have been saying here has any significance at all, it would appear that the very notion of infinity—at least when applied to physical existence, is paradoxical at its very heart. We may indeed speak of infinity in the sense of saying, for instance, that there is an infinite number of points in a line, but points as such are not real physical entities, merely geometric abstractions. Yet even here we cannot completely escape paradox. For instance, two lines of different lengths both contain an infinite number of points. Infinity is paradoxical at its very heart.

## A Universe for All Time?

According to the most recent data from the Planck space probe, the age of the universe is 13.82 billion (1,000 million) years. This is the length of time since, according to the prevailing cosmological theory, the Big Bang set everything in motion.

Now, it is tempting to ask "What happened before the Big Bang?" or "What existed before the time to which cosmologists refer as $T=0$?" or, to phrase it a little more paradoxically, "What was around prior to the beginning of time?"

Many centuries ago, so the story goes, someone asked St. Augustine what he thought that God was doing before he created Heaven and Earth. The Saint replied rather facetiously that God was creating Hell for anyone who asked that question! Nevertheless, the question remained with him and, philosophical thinker that he was, he found that he could not simply pass over it with a frivolous comment. The more serious answer at which he arrived is one that relates the universe and time in a manner that has not been substantially improved upon in the nearly two millennium that have elapsed since Augustine wrote.

Augustine came to see that the question began with a false premise. It assumed that the universe was in time; that time existed in a sense "apart from" the universe and independently of it. If we take the creation of the universe as $T=0$, Augustine's questioner was actually asking what was happening at times before $T=0$, as if time was already flowing before the universe came into being. Augustine saw this as a wrong way of thinking. Time, he concluded, was as much a property of the universe as anything else. Far from the universe existing in time, time exists *in* the universe! As we would say, $T=0$ was not simply the beginning of the physical and spatial universe. It was the beginning of time as well. It really was $T=0$; the beginning of the first moment of time and not simply the beginning of the universe's first moment of existence in time.

The difficulty we have in coming to grips with this probably says more about how we think of time than about any intrinsic weirdness in the concept itself. Stating that "time is in (is a property of) the universe" rather than "the universe is in time" is basically no different from stating that "space is in (is a property of) the universe" as distinct from "the universe is in space". Yet we are not normally tempted to ask "Where is the universe?" or to picture the universe as located in some region of space. We do not find it strange to understand the spatial dimensions as properties of the universe. On the contrary, we find it strange to think of them in any other way. So why should we find it so difficult to apply the same logic to the temporal dimension? Logically, there is no problem, but psychologically, there is.

Paradoxically, confining time to the universe in this way means that it is correct to say both that "There was never a time when the universe did not exist" and "The universe has a finite age". From our common sense point of view, we might conclude that one of these statements must be wrong, yet they each follow quite logically from what has just been said.

Stephen Hawking elaborated on the Augustinian position by drawing attention to a spatial counterpart, namely, that point on Earth's surface which marks the furthest north that one can go; the North Pole. When we say "There is nothing north of the North Pole" we are *not* stating, either, that there is some great void north of the Pole or that some impenetrable barrier exists there preventing further northward travel. Neither are we saying that the North Pole exists infinitely far to the north. On the contrary, the North Pole is simply understood as a boundary point, where the concept of northness reaches its limit and from where every direction on the surface of this planet is south.

A similar boundary exists in the form of a geometric point. Saying that there is nothing smaller than a geometric point does not mean that if we were to shrink beyond the dimensions of a point we would enter into some negative dimension empty of content. It is merely saying that there are *no* dimensions smaller than a point.

With these spatial examples in mind, we can now appreciate a similar boundary situation existing for time as well. $T=0$ is for time what the North Pole is for the direction north on the Earth's surface or for what a point is concerning the dimensions of space. We might even say that $T=0$ is time's point, although the simile should not be pressed too far.

## From Whence Came the Universe?

Traditionally, this has been a question for philosophers and theologians rather than for scientists strictly so called. At the risk of offending some readers, I might say that there were and still are very good reasons for this, but more of that controversial statement in a little while. Let me say first of all that for the sake of simplicity, it is assumed that $T=0$, the absolute beginning of the universe, is co-incident with the Big Bang. That may not be true of course.

There has been considerable speculation in recent times that the Big Bang may not mark the beginning of the universe per se, merely the latest phase in its evolution. The cyclic universe, although now abandoned in its original form, has been resurrected in a different guise as we shall see in the following chapter, and other speculative theories place the true beginning of the universe in an immensely far distant past, seeing the Big Bang as a relatively recent event in terms of overall cosmic history. Nevertheless, even the majority of these theories still propose an absolute beginning of the universe so that the true $T=0$ still took place a finite time ago, albeit longer than the "age of the universe" as derived from Planck data. Such theories are, however, controversial and at the moment lack convincing observational support. The Big Bang, by contrast, rests on strong observational evidence and there seems to be no obvious reason—either theoretical or empirical—to divorce it from the true beginning of the universe. Nevertheless, even if such reasons do appear over the course of time, the philosophical issues raised here are still likely to be relevant, as the issues raised by an absolute beginning are not dependent upon how long ago that beginning occurred.

As it stands, the Big Bang theory is not really a theory about the beginning of the universe so much as a theory about how the universe evolved from an extremely early time. The development of the basic theory to include inflation really only adds more early details without giving an account of what actually happened at $T=0$ itself.

That is not to say that speculation about the cause of the Big Bang is off the menu. Far from it in fact. But there appears to be two questions here which can easily, though not legitimately, fuse into one if we are not careful. First, there is the question of the origin of the spatio-temporal framework in which we exist; that which normally comes to mind when we hear the expression "the universe". The question as to what caused this, what caused the "seed" of the universe to explode into the vast panorama of stars and galaxies that we see today—even what the actual nature of this seed was and how it may have come into being—all seem to be legitimate scientific questions. But then there is the second question, so subtly differentiated from the first that the difference may not be recognizable at first sight. This is the question of why there is something rather than nothing or, expressed in another form, how something could have arisen from nothing. This is the question that has been the traditional field of philosophers and theologians. The domains of the two questions meet at some point, and that point will represent the limits of science. Some readers will no doubt take strong issue with this and insist that nothing falls outside the realm of science. I am well aware that saying the opposite is controversial, but let us take a deeper look at the issue and what is involved.

One idea that has gained a level of popularity proposes that the seed of the universe emerged spontaneously from the quantum vacuum. We have already spoken about the emergence of virtual particle/antiparticle pairs from this restless quantum ocean and we have also seen how real particles of Hawking radiation can be spontaneously created from the quantum vacuum under certain circumstances. Theoretically, just about anything can pop out of the vacuum, yet the more massive the object, the smaller is the probability that this will actually happen. That is why virtual electron/positron pairs are always flashing in and out of existence, but we cannot say the same about boulders, mountains and polar bears. The proposed beginning of the universe as a spontaneous (uncaused) quantum fluctuation is possible in the same sense that a mountain suddenly popping into existence outside my office window is possible. But the probability of such an event is so low that it is in reality highly unlikely to happen. Although some physicists are willing to accept this low probability, others find the proposal quite dubious.

However, this is not the biggest problem with the idea that the universe emerges spontaneously from nothing. The real issue is what is understood by "nothing" in the first place. When someone speaks about the quantum emergence of the universe from nothing, what that person really means is the emergence of the universe from the quantum vacuum. But the quantum vacuum is hardly nothing. It is, as we have said, a seething ocean of quantum fluctuations. It is responsible for the Casimir Effect and the Lamb Shift, each of which represents an observational effect caused by properties of the vacuum. These effects can be looked upon as work performed by features of the vacuum, extracting a certain amount of energy from the quantum vacuum itself. But surely, anything that possesses a store of energy capable of being released as work is not nothing in the true sense of that term. Nothing does not have properties or attributes.

The nature of the quantum vacuum is determined by the Heisenberg Uncertainty Principle, one of the laws of quantum physics. In other words, one of the laws of nature or, phrased just a little differently, one of the laws of the universe. It is not difficult to see where we are going with this: The quantum vacuum empty of material objects, if it ever existed, would have been the universe at that epoch. Therefore, saying that the universe emerged spontaneously from the quantum vacuum really boils down to saying that the universe as we now know it emerged from an earlier phase of the same universe! We are not immediately struck by the tautology because both the words "nothing" and "universe" are quietly ambiguous. The former jumps from meaning absolutely no existence whatsoever (the full meaning of "nothing") to meaning the quantum vacuum (which, if defined as the lowest energy state, is still something in the broadest sense), while the second jumps between meaning simply that which exists (even if *that* is merely the quantum vacuum) and meaning the physical system which is, or which has evolved into, the array of stars and galaxies in which our home planet finds a place.

With respect to the ambiguity of nothing, University of Delaware theoretical particle physicist Stephen Barr used the analogy of the contrast between someone having a bank account with no money in it and that person's not having a bank account at all. On each alternative, the person is insolvent, but the former alternative has certain factors built in so to speak that the latter does not. For instance, if Fred has a bank account, even though he has no money to put into it, there necessarily must exist certain entities called banks. This in turn implies the functioning of a certain type of economic system together with a set of laws and rules governing the operation of banks and banking accounts. It also implies the existence of an already-existing repository that can at any time be used by Fred if or when his financial circumstances take a welcome change for the better.

On the other hand, no bank account at all might mean that Fred is alone on a desert island. It does not imply the existence of banks, an economic system in which these institutions can function or even an economic system at all. In fact, at the most extreme, the reason why Fred has no bank account might be because Fred has never existed or because he once existed but died 100 years ago!

It requires little thought to see the analogy between Nothing equals the quantum vacuum and Fred's empty bank account and between Nothing means complete non-existence and Fred's lack of any bank account whatsoever. If we wished to stretch the analogy a little further, we might even find a correspondence between Fred taking out an overdraft, putting his bank account into the red while acquiring enough cash to purchase a new vehicle, and the emergence of the universe in its present form (or, at least, as something that evolves into this) from the empty bank account of the quantum vacuum. Given Fred's general standing with the bank and a few other factors, Fred could accomplish this, but he would have no hope of pulling money out of a non-existent bank account! (Strictly, the quantum vacuum is not even a completely empty bank account, as we have been discussing above, but if that changes the force of Barr's argument at all, it is only to make it stronger).

This issue can be approached in a slightly different, albeit complementary, way. Science basically explains the universe by discovering the laws by which it functions. These laws form a sort of hierarchy depending upon the complexity of the aspect of the universe being studied. At one extreme, the social sciences such as economics, history and sociology seek the laws by which the complex system of human society functions. Slightly lower down the scale (perhaps?!) are laws of psychology, then biology, chemistry and—finally—physics; the latter dealing with the most basic of physical phenomena. Now, I do not wish to imply a reductionist model in which everything is ultimately reduced to physics. That is going too far. But it is true nevertheless that the laws of physics are still to be seen as the fundamental laws of nature. They ultimately describe the type of universe in which all the other degrees of complexity exist and show in at least a general sense, how these other degrees of complexity can exist.

It has become something of a custom to speak, if not of the laws of nature themselves, at least of the laws governing the disciplines by which these former laws are formulated (that is to say, mathematics and logic) as being laws of thought. That, allow me

## The Shape, Size, Age and Origin of the Universe 139

to object, is a very dubious position to hold. If mathematics and the logical principles upon which it is based tell us only about how we think, why is the world describable in mathematical terms at all? This is not the place to go into this in too much detail, but it is surely the case that the reason why the laws of physics, and indeed most of the laws of the other sciences as well, can be formulated mathematically is because mathematics itself really does describe the way nature works and not simply how we think about it. The rules by which mathematics and logic function only become laws of thought because they accurately reflect in our thought process the way nature functions out there quite apart from our thinking. At the very simplest level, two plus two equals four not because this is how we think about it, but simply because two pairs do in point of fact contain the same number of objects as an array of four objects.

Well, you may ask, what has this to do with our subject? Just this. The laws of nature—including the rules of mathematics itself—belong to the universe. A very different universe logically might have different mathematical rules (we have already seen how geometry differs in a Lobachevskian or a Riemannian universe) and neither mathematics nor physics would have any meaning at all if no universe existed! Like space and time and the rest of the universe which they describe, the laws by which nature functions also came into existence at $T=0$.

The tendency to think of the laws of nature as somehow existing "out there" (but out *where*?) is probably introduced subconsciously by the very term "law". On most occasions when we come across this word it has a prescriptive sense. A law is something which we should obey but nevertheless may also disobey, albeit with unpleasant consequences if we are caught by those whose role it is to enforce it! It is something that is in a sense imposed from above; something that does exist out there, enshrined in legislation, ancient custom, legal precedent and so forth. So when we hear the phrase "laws of nature" (or of physics, chemistry or simply of science) we subconsciously associate this with similar phrases such as "laws of the land" and (also subconsciously) think of natural laws as something pre-existing and imposed upon the universe from outside.

However, the laws of nature are not like that at all. They are descriptive rather than prescriptive. They are not rules which the universe should follow (but may be free to disobey, as we are at least physically free to disobey the laws of the country) but regular patterns that do in point of fact describe how the universe and its constituent parts actually function. It would probably have been better to use expressions like "patterns of nature", "patterns of physics" and so forth, as these phrases have a less prescriptive and more empirical tone to them. Nevertheless, we have what we have and when the laws are written in terms of mathematical formulae, it is tempting to see the formulae as somehow causing the universe to behave in the correctly calculated way. It would appear to be this line of thought that inspired one well known scientist to say that the correct theory of the universe might be so powerful as to have brought about the universe's very existence.

One may argue, however, that there never was a time when there was simply "nothing" in the absolute sense. Maybe there was always a quantum vacuum and at least one law of nature, namely the Heisenberg Uncertainty Principle. Maybe Fred always had a banking account! Following from this, one might argue that the universe as we know it sprang from the quantum vacuum and that time became a property of the universe following this event. On this view, the quantum vacuum is eternal. Indeed, terms like "eternal" and "temporal" hardly even apply to it.

That sounds plausible, but it only works if the quantum vacuum is what philosophers call a necessary being i.e. one that cannot, by its very nature, not exist. That apparently is not the case. Fred Hoyle at one time in his controversial career did exactly that, without any sense that he had made a logical contradiction. But if the vacuum is not necessary, it would seem that something must be logically and ontologically (though not necessarily temporally) prior to it.

Moreover, if Haisch and his colleagues are correct, the quantum vacuum and sub-atomic particles have a mutual interdependence as we saw in the earlier discussion of Haisch's theory of inertia. The mass of particles (according to this thesis) ultimately depends upon the vacuum fluctuations but the latter equally depend upon the zitterbewegung of the particles. A universe sans particles would also be a universe without vacuum fluctuations—literally nothing in the fully fledged sense of that word. Whether this proves to be true or not, the very fact that it can even be stated is enough to show that the quantum vacuum is not "logically necessary" in the full philosophical sense.

If the so-called laws of nature really are descriptions of how the universe does in fact behave rather than prescriptions determining its behavior, it would seem that any hope of explaining the actual beginning of the universe in terms of physical law is a forlorn hope. As remarked earlier, there are good reasons why this has always been the field of philosophers and theologians. Put succinctly, if the laws of nature came into being with nature (the universe) itself, how can the beginning of the universe be explained in terms of these laws? How can the descriptions of how the universe in fact functions be correctly employed to explain how and why the whole thing exists at all? Nature cannot explain nature! The universe is not just weird; it is literally supernatural! Maybe that word carries too much baggage, so how about replacing it with something like transnatural instead? Either way, in so far as science is ultimately the application of the laws of nature to explain natural phenomena, the genesis of the universe and the ultimate explanation for its existence, and for the existence of the very laws of nature themselves, would seem not to be scientific questions.

This statement, I know full well, will not be accepted by all readers. But before you throw this book out the window or expose it to some even less savory demise, let me say immediately that I am not saying that these issues are not the subjects of legitimate inquiry, simply that we should recognize the limits of any theory of physics and be willing to examine the philosophical implications that inevitably arise as physics pushes toward the ultimate matters of cosmology and cosmogenesis.

# The Finely—Tuned Universe; Accident, Design or One-Choice-Amongst-Many?

The more we learn about the universe, the more remarkable it appears but probably the most amazing fact of all to emerge during the past 60 years or thereabouts is the way in which it seems to be minutely adjusted in such a manner as to permit the existence of that phenomenon which we call life. The universe, it appears, is "fine-tuned" for life!

As we discussed in the first chapter, one of the chief attractions of Inflation theory is the manner in which it accounts for one aspect of this fine-tuning, namely, the "flatness" of space. However, that is just one incidence of fine-tuning. Others include the relative strengths of the atomic forces, the ratio of the electromagnetic force to gravity, ratio of proton to electron mass, mass excess of neutron over proton … to name just a few. Considering not just basic properties of the universe at large, but also various specifics of galaxies, stars and planets, over 100 different occurrences of fine-tuning have been noted and new ones continue to turn up all the time.

Neither inflation nor any other theory yet proposed has been able to account for all of these. Indeed, as we have already seen, some physicists argue that inflation has not offered a total solution of the flatness issue either; just pushed the problem back a further step. It is in the consideration of these many examples of fine-tuning that a number of scientists have brought philosophical speculation into the equation.

Broadly speaking, three possible explanations for the fine-tuning of the universe can be proposed. These are;

1) Pure chance. There is no reason for the apparent fine-tuning. It is simply a happy accident. The fact that the universe appears to be so fine-tuned simply relies on the brute fact that if it had significantly different properties, neither we, nor presumably, any curious aliens, would be around to ask these sorts of questions.
2) There are many—maybe even an infinite number of—universes, each with a different set of properties. Most of these will be unsuitable for intelligent life; however a small subset will, quite by chance, possess the fine-tuned set of properties that enable the evolution of creatures capable of contemplating their universe and asking questions about its nature and origin. Ours is one such universe.
3) The universe appears to be fine-tuned because it has been consciously designed for the emergence and preservation of life.

The first of these alternatives, most people will agree, provides an answer that is neither satisfying nor satisfactory. "The universe simply is the way it is, no more questions need be asked" is a reply that simply avoids the issue rather than answers it. Parallel replies would not be accepted in other contexts. For instance, if we asked "Why does hydrogen and oxygen combine to form water?" and was told that "It happens because that is the way nature is" we should rightfully feel cheated. The answer is correct at one level of course, but it completely fails to provide a causal explanation.

It is also quite correct, albeit equally as void of a true explanation, to say that our presence as observers depends upon the properties of the universe being such as to permit the existence of living, conscious, organisms. That is a truism, but it ultimately explains nothing. It does not tell us why the universe is life friendly.

Relating the properties of the universe to the existence of conscious observers is known as the Anthropic Principle. This takes various forms as we shall see. The present formulation is known as the Weak Anthropic Principle (WAP). As we said, it is a truism but may nevertheless be helpful in determining certain parameters of the universe. In a sense and at one level, it could be said to "explain" why we live in a universe with the observed properties, but it provides no true explanation as to why the universe has these properties rather than others. It only explains our observation of these properties; if the universe had other significantly different properties, we would not observe them because we would not be here to observe them!

Another version of the Anthropic Principle known as the Strong Anthropic Principle (SAP) goes a little farther and states that the universe must have the properties that will eventually result in the appearance of life. The exact interpretation of this version rests upon how that little word "must" is understood. In the original formulation of the SAP, "must" was simply deductive—our existence constrains the universe to have the parameters making it suitable for life. That is really not much of an advance on the WAP, but in a more recent formulation by J. Barrow and F. Tipler, the "must" becomes imperative, thereby taking this version of the SAP beyond the present discussion and into the third category of explanation, where more will be said concerning it.

Needless to say, the possibility that a single, chance, universe should be fine-tuned in so many ways as to make life a possibility is extremely remote, which is why the assumption that this just happened strikes us as so implausible. It seems implausible because it is implausible.

This implausibility becomes the starting point for the next proposed explanation. The chance fine-tuning of just one universe is so slight as to be essentially impossible, but suppose that what we call the universe is only one of a large (infinite?) number of universes, each having different laws of physics? Whether we call this a "universe of universes" or a universe of "sub-universes" does not matter. What matters is that the space in which we live—what we have traditionally termed the universe—does not constitute the whole of physical reality. The argument runs that whereas the first alternative assumes just one throw of the dice to produce a fine-tuned universe suitable for life, a large or even infinite set of sub-universes leaves room for a large or even infinite number of throws. Surely, it is argued, a small sub-set of these throws will produce the life-friendly universe in which we live.

This hypothesis gains a certain degree of support from inflationary theory. Inflation per se does not predict a multiplicity of universes, but some formulations of the theory allow for a multiplicity of bubbles of inflating space to form during the very brief inflationary epoch. Our universe may have simply been one bubble, surrounded by others that likewise evolved into different universes with—probably—different durations of their inflationary eras and differing laws of physics.

Nevertheless, this conjecture is not in as strong a position as we might think from the remarks of some of its supporters. For one thing, inflation theory does not uniquely predict multiple universe bubbles. An ensemble of universes is certainly compatible with certain interpretations of inflationary theory, but is not a necessary consequence of inflation per se.

Much will also depend upon the degree of permissiveness allowed within the ensemble of universes. That is to say, how far and wide will the laws of physics range throughout the ensemble? Can these laws range over an infinite set of possibilities?

If the laws of physics studied in our universe are simply the descriptions of the way in which our universe behaves (to express it in a somewhat oversimplified way), it could be argued that even the most basic laws are probably different in any other universes that may exist. Perhaps there are universes where the velocity of light is infinite! Perhaps they are expanding at infinite velocity, in which case they presumably would already have overwhelmed the entire universe ensemble! And why should at least a subset of these hyper-expanding universes not inflate in dimensions other than the three macroscopic ones familiar to us? As will be seen in the following chapter, it is now widely held that even our universe contains extra dimensions albeit manifested only at very small scales, but is that necessarily true for all possible universes?

Some, at least, of the above speculations may be going too far. If we assume that the universe—and all universes—sprang from the same initial state (overlooking, for the moment, the problem of where this came from) there is reason to think that at least certain fundamental laws will be shared and that none of the emerging universes will be *too* weird. But there is no known reason why the inflationary era should last for the same very brief period of time in all universes. Indeed, it is usually assumed that it will not and that this alone will mean big differences in many of the parameters of the different universes. There seems to be no reason why there should not be universes in which inflation simply does not stop. These would just keep on expanding exponentially, overwhelming those sluggish inflators like our own and eventually consuming the entire ensemble! If there is any truth in this suggestion, one should think that our universe would have been "overtaken" long ago.

It is, I think, fair to say that most supporters of the multiple universe hypothesis assume that the laws of physics describing each one will have some basic constraints and that within the framework of these constraints all possibilities will be actualized in some universe. If the ensemble truly is infinite, every possibility will be actualized in an infinite number of universes (but compare this with the earlier discussion concerning an infinite number of O-spheres). The argument runs that even though the exquisitely fine-tuned universe in which we live has an unbelievably low probability of existing, given enough throws of the cosmic dice, it must eventually happen by pure chance. But is that necessarily true? Writing about a slightly different, though closely related, subject (the many-worlds interpretation of quantum physics actually, about which more will be said in the following chapter), Heinz Pagels asked a hypothetical supporter of that theory "how do we know that the other worlds are so different from ours? Maybe different worlds are related to each other like the different configurations of molecules in a gas—each configuration is radically different from most of the others, but what matters is that the gas viewed as a whole looks nearly the same for all those different configurations." (*The Cosmic Code*, p. 178). Pagels' argument, although presented in a somewhat different context as already noted, still holds good in the present case I think. It actually cuts both ways; either all (or most) universes could be fine-tuned like ours or they all could be radically life-unfriendly! Either way, the multiple universe hypothesis fails as an explanation of fine-tuning, but it is probably true to say that, because the chance of fine-tuning is so miniscule, the latter of these two last-mentioned alternatives is the more probable.

The argument that, given enough opportunities, a very unlikely event will occur is one that also needs a closer examination. It is at this point that we can easily fall into what is known as the gamblers' fallacy. This fallacy is not getting mixed up in gambling in the first place (though that is fallacy enough!) but in the assumption that a particular low-probability outcome becomes more likely with each throw of the dice. That is to say, if some result has a 1-in-100 chance of happening, it is a common error to think that if one rolls the dice 100 times, the required result will occur once. The hypothetical gambler acts as though the chance that a low-probability outcome will eventuate becomes greater with an increased number of attempts. To take a simple example, if a coin is tossed, the chances of it coming up heads is 50 % or one chance in two (assuming, of course, that it is not weighted). This remains constant whether the coin is tossed just once or 100 times. At the first toss, the chance of a heads is 50 % … on the one hundredth toss the chance of a heads remains 50 %. Similarly, if 100 coins are being tossed at this very moment in a casino, the chance that my toss will come up heads is also 50 %. The number of other coins being tossed at the same time in no way alters the probable outcome of my toss.

Sometimes, the formulation of the multiple-universes argument seems to overlook this fallacy. If there is a vanishingly small probability that a single universe (one toss of the dice or flip of the coin, so to speak) will come up as fine-tuned in such a way as to be suitable for life, there is an equally small probability that any universe (any toss of the dice or flip of the coin) will. On the other hand, if there is some factor predisposing a universe to turn out fine-tuned, the probability is as great for a single universe as for one of an ensemble, just as a weighted coin has an equally high probability of coming up "heads" (or "tails" as the case may be) on its first as on its hundredth throw. These considerations, however, lead us to the third explanation, but before we go there, a few further words need to be said about hypothetical events of extremely low probability.

Nobody doubts that random events of very low probability do happen. The two unexploded bombs falling at the same spot during the London blitz, mentioned earlier, being a good example. But eventually there comes a limit beyond which the probability of something occurring becomes so low as to be seen as impossible in all practical reckoning. For example, in theory there is a very small chance—somewhere in the order of $1$-in-$10^{80}$ I believe—that the molecules of gas in a bottle will clump together in the bottle's neck, leaving the rest of the vessel a perfect vacuum. But who really expects this to happen, anywhere, anywhen in this universe or in any other? And if it did happen—if somebody brought to a physics laboratory a bottle with a dense globule of air in the neck and the rest perfectly evacuated—the mind boggles as to what might go through the heads of the examining physicists. My guess is that they would accept an explanation involving magic before they would concede that this was the result of a low-probability chance event.

It is impossible at this time to say what the probability of a fine-tuned universe happening by chance really is, because it is unlikely that we know all the factors involved. New evidence of fine-tuning keeps turning up and there is no reason not to believe that even more evidence will be found in the future. But the probability is certainly small, far smaller than anything that would normally pass for a lucky accident.

This brings us to the next hypothesis or, more precisely, to the next set of hypotheses as the suggestion that the universe is purposefully designed comes in a number of packages.

One of these may also be characterized as a version of the multiple-universe hypotheses discussed above. Its supporters accept the version of inflation in which many "bubble universes" form and inflate at differing rates. According to this hypothesis, not all of these form at the moment of the Big Bang but may indeed break away from existing universes at any time. Like cells in a body, universes bud and multiply and even now, our own universe may be begetting children of its own. We would not be aware of this, as once a baby universe forms, it inflates into a space-time continuum of its own and is out of contact with its parent.

A baby universe is hypothesized to begin by emerging from the quantum vacuum of the parent and triggering its own Big Bang, in which its own matter will be created, its very own inflation era and, presumably, its own set of physical laws which (left to chance alone) would most probably be very different from those of its parent. But what if this was not left to chance? If the seed of this process—in a manner of speaking, the embryo of this baby universe—is a quantum fluctuation within the parent, may it not be possible for a highly advanced intelligent race living within the parent universe to both initiate and manipulate the process in such a way that a baby universe having the same fine-tuned physical laws as its parent is born?

This scenario has been seriously proposed as an explanation for the fine-tuning of our universe. We—and by that I mean the entire universe in which we live—might be seen as an artificial creation by a highly advanced alien race living in another, and equally fine-tuned, universe. The mind boggles at the thought that maybe our entire cosmos is the result of a project that earned its creator a PhD (or whatever their equivalent may be!) in some other pre-existing realm.

This speculation would make a good theme for a science fiction novel, but we must admit that it has some serious problems. For one thing, quite apart from the speculative nature of the multiple-universe scenario itself and the difficulties encountered by this model, we simply do not have a clue as to how the quantum vacuum might be manipulated in the manner speculated.

Certainly such a thing is far beyond us and may be intrinsically impossible, irrespective of how advanced a civilization might be.

However, there is an even more basic issue which this suggestion must face. Thus, even if we grant for the sake of the argument that our universe came into being in this way and that this accounts for the manner in which it is fine-tuned for life, the question must be asked as to why its hypothetical parent universe possessed a similar fine-tuning. If this question is answered in the same way—that it too was the result of a universe-creation project in a "grandparent" realm—then the fine-tuning issue is simply passed back one further step ... and so on and on! Either the chain goes back to infinity or the proverbial buck must stop somewhere. Infinity has its problems and most cosmologists would be, I think, rightly skeptical of this sort of infinite regress. But the other alternative is not acceptable either as it means that the original suggestion ultimately fails to explain the very observation (i.e. the fine-tuned life-friendliness of our universe) for which it was put forward in the first place. At best, it simply puts the explanation further back along some hypothesized chain of creation. We still need to ask why the original universe possessed the requisite fine-tuning to bring forth living organisms capable of setting the process in motion. Either this was a sheer fluke (the first alternative) or some intelligence other than an alien in a pre-existing universe was involved. We have already seen the inadequacy of the former, but if the latter alternative is taken seriously, then the chain of universe creation begins to look like an unnecessary complication. It is more economical simply to propose a creating intelligence of a different order for our universe and cut out the chain of middlemen.

Although the subject is a highly controversial one, a number of cosmologists have come to the opinion that the ultimate reason for the existence of the universe is to be found in some intelligence beyond the strict realms of physical science. For instance, after working through the details of the triple-alpha process by which the life-essential elements of carbon and oxygen are synthesized in stars (see Appendix for more detail on this remarkable process) Fred Hoyle remarked that it must be "a put-up job". He found it impossible to believe that such a series of fortunate co-incidences could have been due to chance alone. His developed views on this subject suggest a sort of cosmic hierarchy of intelligent beings in the universe. So advanced are these intelligences that we may as well simply lump them all together under the term "God", although Hoyle did not believe in God in the traditional sense.

An even less traditional and, in the minds of just about everyone who does not subscribe to it, truly weird hypothesis is the so-called Final Anthropic Principle (FAP) put forward by Barrow and Tipler. Following from an interpretation of quantum physics championed by E. Wigner and others (about which more will be said in the following chapter) Barrow and Tipler derived the Participatory Anthropic Principle (PAP) which states that conscious observers are necessary to bring the universe into existence. The full force of this counterintuitive thought will be better appreciated in the following chapter, but for the present it is simply noted as a serious suggestion. It has a certain similarity with the old philosophical chestnut "If a branch falls in the forest and there is no-one or nothing to hear it, can we really say that it makes a sound?"

For Barrow and Tipler however, the PAP merely paved the way for the even more radical FAP. Briefly stated, the FAP says that intelligent life in the universe will continue evolving until it reaches an "Omega Point". Thus far, this line of reasoning appears to run parallel to that of the well known, albeit controversial, priest/paleontologist/philosopher P. Teilhard de Chardin, however the parallels should not be too closely drawn. For Barrow and Tipler the Omega Point is occupied by an entity of such power and intelligence as to have the ability to create its own past! Putting the matter bluntly, what they are saying is that, although there is at present no god, there will come a time in the evolution of the universe and of the life within it when this life will evolve into God, who will then create (back in time) the universe, complete with the fine-tuned parameters necessary and sufficient to enable the evolution of a form of life capable of eventually evolving into God himself. This would give a novel answer to the child's question "If God made me, then who made God?" According to the FAP, God made God!

In his book *The Physics of Immortality*, Tipler argues that evolution toward the Omega Point occurs through advancing computer technology. At present, computer capability doubles about every 18 months or thereabouts so if this doubling rate is extrapolated several millions of years into the future, there will come a time when a computer is created that is effectively omniscient and omnipotent. This computer/god will also, Tipler opines, have the ability to recover all the memories that have resided in the brains of every person who has ever lived and "resurrect" them in a virtual reality!

The immediate reaction is to wonder if *The Physics of Immortality* was first published on April 1. Professional skeptic Martin Gardner speaks for many, I believe, when he suggested that the Final Anthropic Principle be re-named the Completely Ridiculous Anthropic Principle (I will leave it to the reader to work out his suggested acronym!). By way of serious criticism, it must be said that the sort of increase in computer technology envisioned by the FAP exceed the possible limits of computer capability imposed by the laws of physics themselves. Moreover, how is the hypothetical mega-computer of the distant future supposed to collect the memories from brains that turned to dust millions of years before it came into existence? And how is it supposed to turn this collection of data (even if it could acquire it—which it surely could not!) into conscious experience, even if this is within a virtual heaven?

The only possible way would seem to be if this Omega Point Entity was not at the Omega Point at all, but existing alongside the universe from the very start, always knowing everything that happened, anywhere, anywhen, everywhere and everywhen. But such a Being would no longer be the mega-computer of the FAP. Such a Being would be, quite simply, God in the good old-fashioned pre-Tipler sense.

Back in the 1960s, when the present writer was a young philosophy student at Newcastle University in New South Wales, one name that stood out in the God/no-God debate was Anthony Flew. It could almost be said that if atheism had a pope, it was Flew and if it possessed a scripture, it was his *God and Philosophy*. Atheists looked to his writings for support and believers were stimulated by their challenge. Imagine the surprise therefore when, in 2004, Flew announced that he had been mistaken and that he now believed that there is a God after all! Although subsequently expressing a certain sympathy for Christianity, Flew strongly denied that he had experienced any sort of "religious" conversion. His change of mind was purely philosophical and mainly influenced by the complexity of the DNA molecule and the lack of a truly satisfactory naturalistic account of the emergence of something as complex as the first living organism from non-living matter. The more general fine-tuning of the universe did influence his thought to a certain extent and he did name this as a subject that needed to be tackled by anyone debating the existence of God, however it did not form the main thrust of his argument. It appears that what impressed Flew the most was not just the complexity of living things, but what we might call their transitive or teleological complexity, i.e. a complexity that looks beyond the biological entity itself toward some purpose for which it appears to have been designed. In everyday life, things like road signs display this sort of complexity; they are observed to fulfill the purpose for which they appear to have been designed.

In the early 1960s, Fred Hoyle wrote a science fiction story called *A for Andromeda* in which a radio telescope picked up a signal from the Andromeda Galaxy that appeared to have an artificial origin. Subsequently, the signal was decoded and found to include the plans for a highly advanced computer, which the group of scientists who decoded the message constructed—not with altogether good results. The scientists in the story had no doubt that the message was of intelligent origin. Not only was it a complex signal, but that complexity was in the form of a code pointing beyond itself toward some goal, namely, the construction of a very complex device. Although this was pure fiction, it is not difficult to see a similarity between the Andromeda signal and the DNA molecule. Both are very complex and in each instance, the complexity conceals a code giving instructions for the building of a highly complex entity. The biggest difference is that the DNA code gives the blueprint of something too complex to be constructed in the laboratory.

In the opinion of the present writer, considerations of such phenomena as the DNA molecule (upon which Flew concentrated) and the fine-tuning of the universe (our main concern here) are most straightforwardly solved by postulating the existence of God. Of course, one's religious beliefs (or religious non-beliefs) cannot be excluded from this consideration, but the issue itself is not a religious one, as Flew insisted. If the idea of a God seems fantastic, we can only answer that the universe itself is so fantastic that this should not be a concern for us! Moreover, is the thought of an eternal, creative and designing Intelligence less difficult to believe than the self-creating Omega Point of the FAP, an ensemble of universes or a chance happening that is less probable than a lump of rock materializing from the quantum vacuum onto your computer keyboard?

Some may argue that God becomes superfluous if the universe could be demonstrated as eternal. That, however, does not follow. Quite apart from the fine-tuning argument or the points raised by Flew, the issue is not so much whether God is required as a starting point for the universe as whether the existence of the universe itself is what philosophers call contingent or necessary. A contingent being owes its existence to something other than itself. By contrast, a necessary being is one that does not depend upon anything other than itself for existence. As remarked earlier, it cannot by its very nature non-exist! Philosopher Richard Taylor opined that it is easier to understand the concept of a necessary being if we contrast this with the opposite concept of an impossible being, i.e. something which cannot exist—must non-exist, as we might say—because the very concept of such a being is inherently self-contradictory. As examples, he gives square circles and formless bodies. These, by their very nature, cannot exist. The very concept of such things involves a self-contradiction that precludes their existence. As Taylor argued, there is nothing about the universe to suggest that it is necessary in the sense being used here. Every feature of the universe can be imagined not to exist; can be "thought away" so to speak. This is proof enough of contingency; we cannot without contradiction imagine the non-existence of a necessary being, any more than we can imagine the existence of an impossible being such as a square circle.

Yet, try getting your head around total non-existence. Try imagining nothing—absolute and eternal nothing, not even space, time or the quantum vacuum! Try imagining that there is not, never has been or ever will be anything at all! This is literally inconceivable. Just as inconceivable, we might say, as a square circle or a totally formless body. There must, it seems be something; some ground of existence, so to speak. Taylor argues that this contingent universe—the universe consisting of everything that can be thought away—must therefore depend for its continuing existence upon something that is necessary in this radical sense. But if that is true, the relationship between this contingent universe and the necessary being upon which it ultimately depends, remains unchanged no matter how old the universe is; it remains the same, even if the age of the contingent universe truly is infinite. Imagine, says Taylor, that we encounter a flashlight beam. Clearly, the beam is contingent upon the existence of the flashlight that causes it. Yet, if we should discover that the beam has been in existence for all eternity, we should in no way be tempted to doubt that it remained contingent upon the flashlight. We would only conclude that if the beam is eternal, the flashlight must also be eternal. Similarly, if the universe is eternal, all that would imply is that the necessary being must be eternal as well; a conclusion which brings no surprises. A necessary being cannot be other than eternal, for reasons that are quite obvious!

According to Taylor, a necessary being which in effect eternally creates or upholds the contingent universe at least partially defines the traditional concept of God. Combined with evidence of a designing intelligence at work in the basic parameters of the universe, there is reason enough, as Taylor and Flew and others have found, to postulate the existence of God as a hypothesis and not simply as a matter of faith.

By the way, a very similar conclusion concerning the existence of God as "upholder" of an eternal universe was reached by physicist James Coleman of the American International College at Springfield Massachusetts. Coleman argued that the there is a basic misunderstanding in the opinion sometimes expressed that whereas the Big Bang cosmology tends to support belief in God the Steady State cosmology does not. Coleman's argument is essentially the same as Taylor's.

## The Shape, Size, Age and Origin of the Universe 155

By the way, the question "Who made God?" so often raised in this context and assumed by many to be a knock-out counter argument, only has force on the assumption that all beings are contingent. However, if God is a necessary being this question is, in effect, what Gilbert Ryle called a "category error", on a par with the likes of "What color is love?" or, to use one of Ryle's own examples, "You have shown me where the Batsman and Bowler play in the game of cricket, but where does the Team Spirit play?"

Flew and Taylor as well seem to subscribe to a position known as deism, that is, belief in a God who created, set up and maybe even continually sustains the universe, but does not otherwise continue to intervene within it. This effectively becomes the via media between atheism and a fully-fledged theism or belief in a God who is continually involved in the universe or at least in certain of its denizens. The rationale seems to be that natural law is not disturbed by divine (i.e. miraculous) intervention. Personally, I fail to see the logic in this. If physical laws are simply descriptive of the way in which the universe functions and has been set up to function, there seems no reason why there should not be certain laws (albeit not of physics) describing the interaction between God and the universe. Moreover, it does not seem logical for a God of such power and intelligence as to create a universe so fine-tuned as to eventually bring forth lifeforms likewise possessed of a small degree of intelligence, not to be interested in what happens within this universe. After all, a gardener does not normally go to all the trouble of preparing the ground for flowers or fruit trees only to then sit back and let the weeds take over! Moreover, even though we might not go along with Tipler's version of the Omega Point, it is surely not unreasonable to believe that a Creating Intelligence would have something not too unlike an Omega Point in mind; something other than an ultimate empty void into which the universe will eventually sink. As already mentioned in passing, perhaps life, intelligent life, will eventually come to a point where it plays a role that we can scarcely conceive of at present. If so—or if life has some preordained role that we cannot even comprehend today—of course God would be interested in its progress. Ancient faith got it right!

Nevertheless, to pursue this any further would be to stray too far from our topic. We shall now turn from the grand overall picture of the universe and look into the very small; the ultimate components from which the universe is made. It is here that common sense images are left far behind and, by popular agreement, the weirdest aspects of our weird universe are to be found.

# 4. Of Atoms, Quanta, Strings and Branes; Just How Weird Can the Universe Really Be?

Every schoolchild knows about Isaac Newton, the great seventeenth century physicist whose theory of universal gravitation made it possible to calculate everything from the motions of the planets to the trajectories of rockets ... or falling apples according to the popular story. But Newtonian physics had an even wider effect on scientific thought. His model of the universe was that of a great mechanism; a machine functioning according to strictly deterministic laws with the precision of clockwork. The Newtonian universe was, in short, a clockwork cosmos. What is now referred to as classical physics grew out of this conception of nature. Basically, physicists of the post-Newtonian era saw themselves as investigating the laws by which this great clockwork mechanism operated and refining them to ever greater precision.

To a large degree, this clockwork model served the scientific community well. Great advances were made in, for instance, the field of celestial mechanics. The discovery of the planet Neptune stands as a spectacular example as to how calculations using Newtonian gravitational theory could predict the presence of something as yet unseen. Once the orbit of the "new" planet Uranus was found to slightly diverge from its predicted Newtonian path, it was realized that some unseen mass must be perturbing it—some as-yet unseen wheel of the clockwork mechanism must be present—and through the employment of the same gravitational theory, the position of this undiscovered body was computed and subsequently verified by observation. A host of other discoveries were also made within the framework of the Newtonian worldview—thermodynamics, James Clerk Maxwell's electromagnetic wave theory of light, all fitted quite neatly into the prevailing worldview of a strictly deterministic universe.

Indeed, in his book *Modern Science and Materialism*, written early last century, Hugh Elliot could boldly state that if we knew all the laws of physics and had total knowledge of the distribution of matter within the universe, it would be possible to determine even the thoughts inside people's heads. Of course, that could never happen in practice, but even stating the theoretical possibility is enough to demonstrate the rigid deterministic nature of the Newtonian world. As Heinz Pagels wrote:

> [within the Newtonian universe] All that happens – the tragedy and joy of human life – is already predetermined. The objective universe exists independently of human will and purpose. Nothing we can do can alter it. The wheels of the great world clock turn as indifferent to human life as the silent motion of the stars. In a sense, eternity has already happened. (*The Cosmic Code*, p. 19).

This deterministic world was also a continuum. Sharp and discrete distinctions were looked upon with suspicion. Nature did not draw distinct lines but appeared to favor a blending of different forms of matter one into the other. This feature of the natural world was apparently not confined to what might be called inanimate nature, but applied even to biological species as demonstrated by Charles Darwin's writings. How far the division of a piece of matter could proceed was not agreed upon, although the broadest consensus amongst physicists was that there was a limit to divisibility in the form of tiny indivisible particles known as atoms. This was a theory that had been around ever since Leucippus and Democritus in ancient Greece, but even by the late nineteenth century had not succeeded in winning over all scientists. One such skeptic was philosopher-physicist Ernst Mach, who argued that anything supposedly beyond the verification of experimental evidence could not be claimed to exist. In Mach's opinion, because atoms could not be verified empirically, they could not be said to exist. Given that he was a convinced Machian in his early years, Einstein may initially have thought that way as well.

Yet, certain phenomena did not fit well into this well-oiled clockwork continuum. The discovery of radioactivity presented a potential problem. Why do some materials spontaneously emit particles and rays? From where do these emissions come and why is their appearance so spontaneous, so random? And then there were the difficulties raised by spectroscopy. The continuous rainbow spectrum from red through to blue was not a problem, but what was to be made of the sharp lines observed in the emission spectra of different substances? These presented a real mystery which the physics of the time—the late nineteenth century—was powerless to solve.

The revolution in scientific thinking that was destined to profoundly change our view of the natural world from a strictly regulated to a fascinatingly weird place began in the year 1900. Like most of the revolutions that actually do bring about fundamental change and are therefore truly worthy of the title, this one began quietly and inauspiciously.

The spark that fired the revolution was the problem of blackbody radiation. Imagine a cool bar of metal kept in a totally dark room. Not being luminous and being in an environment free from any incident light that it might reflect, this metal bar is totally invisible, totally black. Next, imagine that the bar is briefly taken from the room and heated until it starts to glow. When returned to the dark room, it is now visible as a dark-red luminous object. If it is once more taken from the room and heated to a very high temperature, upon its return to the otherwise dark room it will now be visible as a source of white light. The light emitted from the hot body in the dark room has a distribution of colors capable of being measured and graphed in what is known as the black-body radiation curve.

Two teams of experimental physicists at the Physikalisch-Technische Reichsanstalt in Berlin had made precise measurements of this black-body radiation curve and on the basis of their results, physicist Max Planck attempted to understand the phenomenon in terms of the theory of heat. At first, he did not seem to be getting very far but then in what we might like to call a flash of intuition (but which he actually described as "an act of sheer desperation") Planck abandoned the accepted view of the radiating black body as a lump of continuous matter in favor of treating it as a swarm of vibrating oscillators whose energy exchange with the black-body radiation was quantized i.e. consisted of a myriad of discrete energy packages. The energy emission was not

FIGURE 4.1 Max Planck 1858–1947

continuous as classical post-Newtonian physics and common sense believed, but (in a sense) atomized. The continuous appearance was somewhat on a par with the apparently continuous appearance of a pile of rubble. From a distance, a rubble pile looks like a single continuous heap of material. Only upon closer inspection is it found to consist of many small and discrete individual fragments of stone (Fig. 4.1).

Planck's act of sheer desperation was indeed a stroke of genius, although neither Planck nor anybody else at that time could have foreseen where it was to lead.

Planck denoted the numerical measure of this degree of discreteness by the letter $h$, later called Planck's constant in his honor. Referring to our example of the rubble pile, $h$ is analogous to the size of the individual pieces of rubble comprising it. If the value of $h$ is 0, we are back to the old picture of a continuum, but if it has a positive value—as experiments soon revealed to indeed be the case—the appearance of continuum gives way on closer inspection to an ensemble of discrete or quantized packets of energy.

In retrospect, but only in retrospect, Planck's discovery opened the door to a very different universe from the one in which the physicists of 1900 thought that they lived. But the door was only opened by a crack initially. The classical world of clockwork and continuum was undisturbed except for a little hint of weirdness around the edges. Physicists not immediately involved with black-body radiation paid scant heed to Planck's insight.

Then, just 5 years after Planck's act of sheer desperation, a young doctoral student working as a patent clerk to pay for his study published three very important papers. The young man was, of course, Albert Einstein and the three papers dealt with his favorite subjects of statistical mechanics, the photoelectric effect and special relativity. Remarkably, these three epoch-making papers were published while he was still technically an amateur physicist.

In his first paper, Einstein sought to provide evidence for the existence of atoms through observational means. We earlier wondered if Einstein may have been an atomic skeptic very early in life but whether that was so or not, it seems clear that by 1905 he was convinced of their existence. Yet, true to his mentor Mach, he was not satisfied with anything less than empirical proof that such things were real. This he proposed to find through observation of very small particles in a colloid. If the particles were small enough—around 0.001 mm in diameter—impact by atoms and molecules (themselves composed of atoms) should cause them to jiggle slightly and this effect should be large enough to observe under a microscope of sufficient power. Indeed, just such a movement had already been noted for tiny particles suspended in a liquid. The phenomenon was first noticed by Scottish botanist Robert Brown who observed a mysterious jiggling of very small particles suspended in the medium within pollen grains (not the pollen grains themselves) but with no explanation forthcoming, this had remained just another of life's mysteries. Einstein however, made quantitative predictions based upon atomic theory, for the magnitude of the effect and these were verified experimentally by J. Perrin a few years later. (As an aside, although this convinced most scientists—even former skeptics—of the reality of atoms, Mach remained unmoved in his skepticism for the rest of his life).

## Project 5: Brownian Motion

Several techniques have been used to demonstrate the presence of Brownian motion, including special chambers for detecting it in suspended smoke particles. The following is arguably the simplest method and comes with the recommendation of David Walker who has employed it to show Brownian motion in minute fat globules within a drop of very dilute full-cream milk. If you own, or have access to, an ordinary student's optical microscope capable of 200, or preferably 400, magnification, you may like to try it for yourself.

First, place a small drop of water on the microscope slide. The water used should be as pure as possible, preferably distilled or, in the absence of this, at least boiled or collected directly from the melt of clean ice.

Next, dip a needle, first into some full-cream milk and then into the drop of water, stirring around to ensure that the trace of milk is well mixed and well diluted (undiluted milk would be too crowded with fat globules).

Then, gently place the coverslip onto what is now a drop of very dilute milk, making sure that none of the liquid creeps out from around the edge of the coverslip (as this will produce unwanted currents in the solution as the exposed liquid evaporates). It helps to seal around the edges of the coverslip, which may be done by using a needle to lay a thin continuous line of petroleum jelly along the coverslip edges.

You are now ready to place the slide under the microscope. Illuminate the slide in the normal way and use lower power to locate the fat droplets within the solution. Then change to high power and check the droplets for motion.

At first, the fat droplets may be moving across the field of view. This is not Brownian motion. Rather, it is caused by the presence of small currents in the liquid under the coverslip. If this type of movement is seen, wait for a few minutes until the liquid settles down into a steady state and the droplets stop drifting around.

> When they settle down, examine them carefully under high magnification and note whether a slight jiggling motion is present. This is Brownian motion caused by the bombardment of the fat globules by molecules of water. Note whether there is any difference in the degree of "jiggle" between fat globules of different size. Where is the jiggle most apparent?

Einstein's second paper dealt with the mystery of why, when a beam of light is shone onto a metal surface, electrically charged particles or electrons are emitted by the metal, resulting in a current of electricity. It was here that Einstein relied on Planck's findings of 5 years earlier. Einstein asked why, if heat energy comes in discrete packages, should the same not also be true of light? He saw no reason why that should not be so and proposed the very radical idea that light itself consisted of quanta or photons. When a ray of light is shone on a metal surface, what actually happens is that the metal surface is bombarded by a hailstorm of photons and it is through this bombardment that electrons are released.

This idea won little support, so deeply ingrained was the concept of light as a wave. Even Planck did not believe it, and there is evidence that Einstein himself came to reject it. Ironically, the particulate concept of light was one that Newton had proposed, so it appeared at both the beginning and end of the period of classical physics. It was eventually confirmed experimentally in the early 1920s.

Einstein's third paper of 1905 was his celebrated work concerning Special Relativity, about which we earlier spoke.

## Solar-System Atoms and Quantum Leaps: Rutherford & Bohr

Within 6 years of Einstein's paper confirming the existence of atoms via the observation of Brownian motion, new and surprising information as to the structure of these particles was to emerge through a series of experiments conducted at Manchester

164　Weird Universe

FIGURE 4.2 Ernest Rutherford 1871–1937 (modified from original, *credit*: Diego Grez)

University by Ernest Rutherford and his assistant Ernest Marsden. Rutherford was studying the alpha particles (nuclei of helium) that are emitted by certain radioactive substances. He rigged up a sort of benign ray-gun, with a radioactive source emitting a stream of alpha particles directed toward a screen coated with a substance that emitted small points of light each time an alpha particle struck it. A strip of gold foil was placed between the source and the screen, such that the alpha particles needed to pass through the foil before reaching the screen. Most of the particles passed directly through the foil with no disturbance; however a small number suffered significant deflection. As the screen was moved further from the direction of the main particle stream, occasional flashes continued to be seen. Clearly, some particles were strongly deflected, the odd one even bouncing back a full 180° to the source itself! (Fig. 4.2)

Because alpha particles are positively charged, Rutherford concluded that the deflected ones must have passed very close to the positive nuclei of the atoms in the foil and been repelled by the close encounter with a like charge. From this, and from a series of experiments that his early results initiated, he announced

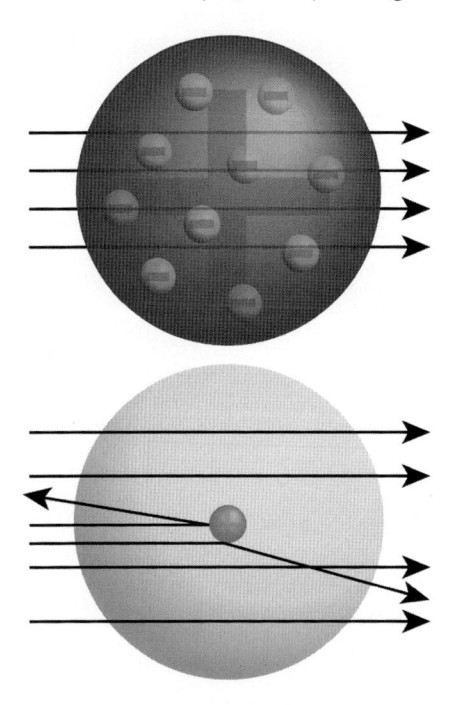

FIGURE 4.3 Rutherford's gold foil experiment. *Top image* is what was expected to happen; *bottom image* depicts actual observations (*Credit*: Kurzon)

in 1911 a new model for the atom. In this model, an atom was pictured as consisting of a tiny positively charged core or nucleus, in which most of the atomic mass resided and around which orbit negatively charged electrons. The atom would appear—if we could actually see it—like a miniaturized solar system and (just like our actual planetary system) would consist mostly of empty space (Fig. 4.3).

As happens quite often in science, this new model solved the problem at hand, but raised further difficulties that needed to be tackled.

The chief problem with Rutherford's atomic model was its stability. According to classical physics, moving electrons radiate a continuous emission of electromagnetic radiation. But if that was really happening to the orbiting electrons of Rutherford's model, they must inevitably lose energy and over time spiral down into smaller and smaller orbits. An eventual merger with the atomic nucleus seemed inevitable. So why aren't atoms discrete,

FIGURE 4.4 Niels Bohr 1885–1962 (*Credit*: Nobel Prize Biography, A. B. Lagrelius & Westphal photo)

solid and electrically neutral micro-particles; combinations of positive nuclei with the electrons that once orbited them now fused into their miniscule masses?

Then there were those pesky bright lines seen in emission spectra. One would think that as electron orbits decayed and fell into lower energy levels, their emitted wavelengths should become increasingly longer and any emission lines so broadened and blurred as to essentially merge into a continuum. Clearly, that it not what is found.

About a year or thereabouts after Rutherford published his results, a young Danish physicist by the name of Niels Bohr came to spend about 6 months with Rutherford between his leaving of Cambridge and return to his native Copenhagen. This brief collaboration with Rutherford was destined to precipitate some remarkable results (Fig. 4.4).

Bohr realized that the problems raised by the Rutherford atomic model could not be solved within the boundaries of classical physics but fell away readily if attacked through the quantum framework laid down by Planck and Einstein. With the quantum

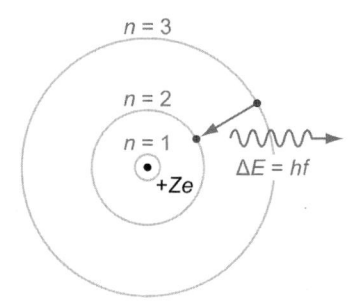

FIGURE 4.5 Bohr's model of the atom (*Credit*: Jabber Wok at the English Language Wikipedia)

structure of light and energy in mind, Bohr made the simple assumption that the electrons surrounding an atomic nucleus did not continually radiate light (as classical physics would have us believe) and that only specific orbits for electrons are allowed. In other words, if energy is quantized, electrons cannot spiral continuously into lower orbits. If they must change orbits, they can only do so in quantum jumps. The energy released during these quantum jumps is itself quantized. It is emitted in light quanta or photons and, because only certain electron orbits are allowed for specific atoms, the energy of these photons is specific. Atoms can only emit light at characteristic and very specific energy levels (i.e. colors), hence the existence of those erstwhile mysterious sharp lines in emission spectra and hence their positions in the spectrum characteristic of atoms of various elements.

Still, just as Rutherford's atomic model solved some problems at the price of raising others, so Bohr's quantum model did likewise. Questions were immediately raised concerning the cause of the jumps, the direction of the emitted photon and why that direction should be taken rather than another. There seemed, in short, to be no reason why any of this happened. These atomic processes appeared to be happening spontaneously. Einstein, die-in-the-wall determinist as he was, was troubled by this apparent spontaneity, but he also recognized that a similar spontaneity took place in the phenomenon of radioactive decay. Whether Einstein liked it or not (and it is abundantly clear that he did not!), the old deterministic clockwork universe of classical physics was starting to crumble; at the level of atomic phenomena at least (Fig. 4.5).

## The Atomic Matrix: Heisenberg & Born

Following his brief collaboration with Rutherford, Bohr returned to Copenhagen where he soon made a name for himself as a physicist. So much so, in fact, that in 1920 several Danish businesses, including the Carlsberg brewery, donated sufficient funds to allow him to realize his dream of founding an institute that would draw the cream of physics students from many countries specifically to work on the issues raised by atomic theory. The following year, a bright young German named Werner Heisenberg heard Bohr lecture on the latter's favorite subject and, through subsequent discussions with the man himself, decided to steer away from his original choice of pure mathematics and toward the study of atomic physics. Heisenberg spent a year at the Niels Bohr Institute (as the establishment had by then became known) before leaving for several years to study under Max Born at the physics institute of the Gottingen University. It was there that Heisenberg hit upon a truly revolutionary idea that was to lay a further building block in the emerging structure of quantum physics (Figs. 4.6 and 4.7).

FIGURE 4.6 Werner Heisenberg 1901–1976 (*Credit*: Bundesarchiv, Bild 183 – R57262/CC-BY-SA)

FIGURE 4.7 Max Born 1882–1970

Heisenberg began to approach the problem of atomic structure, not in terms of pictorial imagery as to what an atom might look like (miniature solar system or whatever) but in terms of what an atom did; in terms of its energy transactions. Obviously still a mathematician at heart, he approached the subject mathematically, describing the energy transactions of an atom by way of an array of numbers. His next step was to discover the rules that these arrays of numbers obeyed and then to use these in the calculation of atomic processes.

Heisenberg moved back to Copenhagen after 4 years or so, but before leaving Gottingen, he showed his work to Born who immediately recognized his number arrays as matrices. Born and his student Pascual Jordan worked further on Heisenberg's ideas and published a paper describing how matrix algebra seemed to be the way in which atomic energy transactions can be described.

This, once again, broke with classical physics. In classical physics, both the position of a particle with reference to some fixed point and that particle's momentum are described in terms of simple numbers. Moreover, as every schoolchild knows, simple numbers obey the commutative law of multiplication. In other words, if A and B are simple numbers, then;

$$A \times B = B \times A$$

Therefore, according to classical physics, variables such as the position and momentum of a particle, being capable of denotation as simple numbers, must also obey this law.

However, according to the matrix mechanics discovered by Heisenberg, Born and Jordan, these variables are not simple numbers but, instead, are matrixes or arrays of numbers. Unlike simple numbers, matrixes do not necessarily obey the commutative law of multiplication. In other words, $A \times B$ may not be equal to $B \times A$ if both B and A denote matrixes. Specifically, Born and Jordan worked out a relationship between the position (p) of a particle and its momentum (q) such that the difference between $p \times q$ and $q \times p$ is proportional to Planck's constant. Only if the latter is equal to zero (implying that energy is continuous and not quantized) would $p \times q = q \times p$. But since it had already been amply demonstrated that Planck's constant is always positive, this equivalence does not occur in the real world.

As an aside, it is interesting to note that Heisenberg does not appear to have been acquainted with matrix algebra at the time he worked out his mathematical model of the atom. It seems that he virtually discovered this field of mathematics as he wrestled with a problem of physical reality, viz. atomic structure. This recalls another example of the close connection between physics and mathematics (albeit coming from the other direction, as it were) when Paul Dirac hypothesized the existence of an anti-electron on the grounds of the mathematical law that the square root of a positive number has both positive and negative solutions. The anti-electron (subsequently discovered and now known as the positron) corresponded to the negative root. This once more demonstrates how the laws of mathematics, like the laws of physics, are basically descriptions of how the natural world functions and not primarily how we formulate our thinking about the world.

Dirac, we must also note, played an important role in the development of the revolution of thought that Heisenberg's insight unleashed. Hearing Heisenberg deliver a guest seminar at Cambridge late in 1925, Dirac was so impressed with the new theory that he soon afterwards published a paper formulating matrix mechanics as a complete dynamical theory replacing classical mechanics.

FIGURE 4.8 Wolfgang Pauli 1900–1958 (*Credit*: Fermi National Accelerator Laboratory)

Soon thereafter, the brilliant and fiery young physicist Wolfgang Pauli demonstrated that the new matrix mechanics was equally efficient as Bohr's atomic model in explaining spectra and indeed went beyond this by accounting for problems associated with the spectrum of hydrogen in an electric or magnetic field; something which had not been explained earlier, not even by Bohr's model (Fig. 4.8).

Clearly, this new approach was a powerful one, but it did leave one big philosophical problem in its wake. Granted, it told us a lot about what the sub-atomic realm does, but does it actually tell us what the denizens of this tiny realm are like … really like?

Heisenberg and Dirac denied that this even constituted a valid question. In their opinion, trying to think up models as to how atoms look (or would look if we could somehow see them) is a dead end. If we are to understand atoms, the only way in their opinion, is via the discipline of mathematics.

This approach did not leave everybody laughing however! Quite the contrary, many physicists were not at all happy with such a reply.

FIGURE 4.9  Louis de Broglie 1892–1987

One such was Louis de Broglie. De Broglie was impressed by Einstein's theory of light as consisting of quanta and his radical suggestion, made in 1909, that because there was equally strong evidence that light was also a wave, a future theory of light would need to fuse the particle and the wave theories into a single model. What de Broglie proposed was that such a fusion of models not be confined to light, but that it should also be extended to include electrons. Now, although electrons clearly acted as particles in certain circumstances, there was clear evidence that they behaved like waves in others. Maybe, just as Einstein proposed for light, they indeed were both! (Fig. 4.9)

Einstein himself was impressed with this idea. So was Austrian physicist Erwin Schrodinger, who further developed the atomic model by including the Janus-faced electron concept and demonstrated that it had the same predictive power as the earlier models of Bohr and Heisenberg. Schrodinger proposed that an electron is a matter wave which nevertheless can be approximated as a particle. Moreover, he argued, this applies to all quantum objects, not just electrons and photons. As against Heisenberg and Dirac, Schrodinger insisted that an atom can indeed be pictured conceptually and not only as a mathematical construction.

FIGURE 4.10 Erwin Schrodinger 1887–1961

It can be conceived of in terms of real, physical, waves. De Broglie even used this model of physical waves to predict that sub-atomic particles would display phenomena usually associated with waves, such as interference patterns. This prediction was verified in 1927. The physical wave interpretation was later developed by David Bohm and seemed about as close to a common-sense interpretation as the increasingly weird world of quantum physics could manage. At least, it is close to common sense except for one profoundly non-commonsensical fact. According to the mathematics of the wave function, the waves themselves must occur, not in our familiar three-dimensional space, but in a hyperspace having many dimensions!

In the face of this difficulty, what we might call the physical wave model, was not the one that became widely accepted, despite its obvious success in presenting an apparently viable atomic model (Figs. 4.10 and 4.11).

Recall that Heisenberg's matrix mechanics precludes the properties of quantum particles, for example their position and momentum, from being determined at the same time. Imagine a quantum particle—an atom or free electron for instance—flying through space. Because of this uncertainty, its exact path cannot be predicted

FIGURE 4.11 David Bohm 1917–1992

unlike, say, a flying cricket ball in the familiar world. But by applying Schrodinger's wave function (just accepting the mathematics and not worrying about whether his material interpretation is correct) one can nevertheless work out the probability of finding the quantum particle at any given position. Thus, the wave function gives probability and from this Bohr and Heisenberg concluded that what Schrodinger's wave calculations are dealing with are probability waves, not real ones in the sense of literal waves propagating through some material medium. Maybe these scientists found it easier to believe in something as abstract as probability waves than actual ones requiring a multitude of extra dimensions.

The subtle difference between the Schrodinger and Bohr/Heisenberg approach is summed up in this simple analogy given by Marcus Chown in "Ghost in the Atom" published in *New Scientist,* July 2012;

> Say you arrive at a lake into which a large number of plastic bottles have been thrown. You notice that there are places where the bottles bunch up and places where there are few bottles. By counting the number of bottles at different

locations, you could create a probability distribution, which allows you to estimate the chance of finding a bottle at each point.

But suppose you notice that the bottles are most common where the amplitude of the real waves in the lake peak. Now you realize that the probability distribution is not the last word – there is a mechanism behind it. Real, physical waves have driven the bottles to their particular locations.

The first paragraph of this quoted passage is analogous to the Bohr/Heisenberg approach and the second to that of de Broglie and Bohm, following on from Schrodinger. Both alternatives at least agree on one thing—the universe at the quantum level is truly weird. Probability waves or physical waves existing in a multi-dimensional hyperspace; those are the choices. Either is beyond our capacity to simply picture in terms that would make sense to minds conditioned to function in a macroscopic classical reality.

# What Does It All Mean?

All of the remarkable insights from Planck and Einstein to Bohr, Heisenberg, Schrodinger and the rest of the boys in the band were tributaries merging into the great confluence of quantum physics. This has been remarkably successful and widely practical. Much of modern technology, including the very computer being used by the writer, depends upon its insights into how the quantum world works.

But what does it really mean? That was the question from the very beginning and it remains so to this day.

Back in 1926 when Erwin Schrodinger first put forward his theory of wave functions, he denoted the wave function by the Greek letter psi ($\psi$), which inspired one fellow physicist to pen the following;

> Erwin with his psi can do
> Calculations quite a few
> But one thing has not been seen:
> Just what does psi really mean?

As we have seen, it meant one thing to de Broglie and Bohm (and Schrodinger himself) and quite something else to the likes of

Bohr and Heisenberg. And, which will hopefully become clearer in a little while, something else again to many interpreters of Bohr, both professed supporters and critics.

The big problem is that while the wave function actually encodes within it the many possibilities for a quantum particle's state, we can in actual fact measure only one of these. What then, does this say about the other possible states—are they equally real? Or perhaps none of the states is *really* real? At least, not until it is actually measured. But does that mean that we—the observer(s)—somehow create(s) the observed state?

The probability wave interpretation of Bohr and Heisenberg appears to imply something along the lines of the last alternative. It seems to be saying that, to take one example, an electron can only exist as a real particle at a point in space where we can observe it directly. In experiments in which it acts as a wave, it is a wave; in those where it acts as a particle, it is a particle. That does seem to have been Heisenberg's view. Whether it was Bohr's will be taken up a little later.

If this position should be true, the question can be raised as to whether or to what extent this quantum indeterminacy spills over into the macroscopic world. Does it mean that there is no such thing as an objective world? Forget about the question of whether a branch falling in the forest makes a sound if nobody hears it. Following this strict subjectivist line that Bohr supposedly opened up with his interpretation of the Schrodinger wave function, we might now need to question whether we can even speak about the branch falling at all unless someone perceives it. Some of the more extreme interpretations imply that we can only speak about the branch having fallen if and when it is observed to be in the fallen state. If a hiker, walking through the forest today finds a branch that fell from a tree yesterday, his perception of this fact is somehow supposed to reify (we can hardly say cause) yesterday's fall. Until then, the branch exists in a fallen/not fallen Janus state. Lest the reader might think this a parody, it might be timely to mention a recent article in a respected science journal suggesting that the recent discovery of dark matter may have changed the life-expectancy of the universe. Before it was observed, the dark matter did not really exist and therefore played no role in cosmic evolution!

Such far-out ideas were not presented to Schrodinger, but he apparently saw the way that things were going when his matter waves became replaced by probability waves. This led him to construct a thought experiment which he apparently believed to be the reductio ad absurdum of the subjectivist interpretation which he saw developing from Bohr's position. The thought experiment involved what has become one of the most famous fictional cats in history, next only to Felix and Garfield! Ironically, so convinced have many people become that everything about quantum physics is ultra-weird, that what Schrodinger saw as an obviously absurd conclusion has been taken on board as another of the oddities of the real universe that quantum theory has brought to light!

## Schrodinger's Cat

Schrodinger imagines a cat confined in a sealed container within his laboratory. Because the container is so securely sealed, there is no way that anyone outside the container can know what happens within it. Now, together with the cat, there is a weak radioactive source and a detector of radioactive particles, which in turn is connected to a trigger capable of smashing a phial of poison gas. The radioactive source is such that there is a 50 % chance that a radioactive particle will be emitted during the time span of one minute. The detector is turned on for exactly one minute, giving it a 50 % probability of detecting a particle. If a particle is registered, the trigger will be sprung, the phial broken and a cloud of gas released with fatal consequences for the poor old cat! (Fig. 4.12)

According to the probability wave interpretation, there is an equal probability that a particle will or will not be emitted, so there is an equal probability that the cat is alive or not alive at the end of the minute. Schrodinger argued that, according to Bohr's interpretation, a probability wave can be assigned to the state of the dead cat and another to the state of the living cat. The situation inside the box is therefore a wave superposition state consisting of an equal measure of the dead-cat probability wave and of the living-cat probability wave. In short, the cat is neither dead nor alive but is described as a superposition of both states. It remains in this state until the box is opened and its content (either a dead cat or a living cat) is observed. In a sense, the observation places the cat in

178  Weird Universe

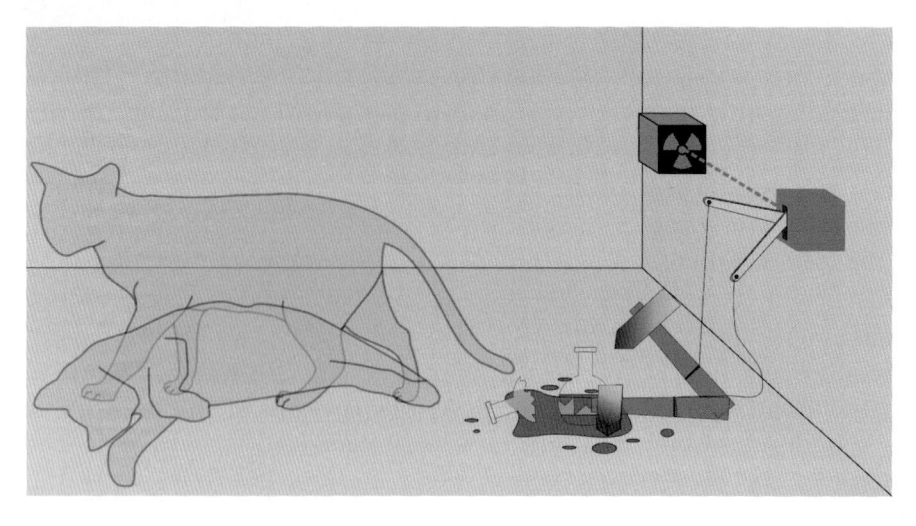

FIGURE 4.12 Schrodinger's Cat thought experiment. Is the cat dead, alive or somehow in both states before the box is opened and the contents observed? (*Credit*: Doug Halfield)

one or other of the two equally probable states. By all common sense standards, this seems absurd.

Moreover, even if we grant, for the moment, that observation is necessary to bring the cat into one or other state, we must ask at what level this observation is to be placed. By this I mean, prior to the opening of the box, no-one knows (at least, no human knows) the result. Then the scientist in the laboratory knows, but what if the laboratory itself is sealed? For those in the wider world, the outcome of the experiment is still as mysterious as it was while the cat's box was sealed.

This line of thought led another physicist, Eugene Wigner to propose an upgraded version of Schrodinger's thought experiment which, though not as well known as the original, raises some very intriguing speculations.

## Wigner's Friend

Wigner pictures a similar setup to Schrodinger's original; a sealed chamber, a weak radioactive source having a probability of 50 % emission of a particle during the span of one minute and a detector capable of registering any particles that may emerge.

FIGURE 4.13 Eugene Wigner 1902–1995 (*Credit*: Nobel Foundation)

More humanely however, this time the detector is not connected to a poison gas source, but takes the form of a far simpler and more benign screen coated with a material that fluoresces in a flash of light when struck by a particle. Instead of a cat, Wigner imagines a human assistant inside the chamber. As before, the detector is timed for precisely one minute, during which the person in the chamber maintains careful watch on the screen, ever vigilant for the possibility of that fleeting flash of light (Fig. 4.13).

At the end of the minute, the chamber is opened and Wigner's friend emerges. Wigner asks her whether she observed a flash during the duration of the experiment and she answers either "Yes" or "No". So just like Schrodinger before him, Wigner now has knowledge of what happened within the chamber during the time that it was sealed and its contents unobservable from the rest of the world. But does it therefore follow that before Wigner opened the chamber, his friend was in a state of Janus-faced reality like Schrodinger's cat? Not on this occasion in a dead/living dual state, but in a superposition of having-seen-a-flash/not-having-seen-a-flash dual states?

Wigner thinks not. Whilst he is willing to admit that a purely mechanical device (or a cat?) registering such an event might be described in this way, the presence in his thought experiment of a conscious observer complicates the matter in so far as his friend's senses communicate the information to her consciousness at the time of the experiment, even though she is prevented from giving this information to Wigner himself, and to the wider world if required, until the later time of the chamber's opening. Wigner then concludes that "the being with a consciousness must have a different role in quantum mechanics than the inanimate measuring device." He proposes that a conscious experience be understood as an objective reality that is correlated to a change in an objective, albeit probabilistic, wave function. What we may call our knowledge is then understood as the aggregate of the conscious knowledge of all systems possessing conscious awareness. In this way, he hopes to characterize quantum physics as an objective theory describing the interaction between an objective physical aspect describable in terms of the mathematical constructs of quantum mechanics and an equally objective mental aspect describable in terms of psychological concepts such as thoughts, ideas and the like. Consciousness is located right at the heart of the system.

This sort of elevation of human consciousness is what many New Age types like to hear and the promotion of this as orthodox quantum physics appears to give it a respectability that it otherwise might not have. But is this really as orthodox as it is often said to be?

## How Did Bohr Interpret Quantum Reality?

If we each had a dollar for every time the expression the Copenhagen interpretation of quantum physics is used, we would all be wealthy people! Moreover, as the founder of the Niels Bohr Institute, where the Copenhagen group conducted its research, it is frequently taken as axiomatic that what has since become known as the Copenhagen interpretation was originally Bohr's interpretation and that the rest simply followed it. It is the orthodox interpretation of Quantum physics. And lastly, it is presented in terms of radical subjectivism.

It might come as a surprise therefore to read the following; "until Heisenberg coined the term [i.e. "the Copenhagen interpretation"] in 1955, *there was no unitary Copenhagen interpretation of quantum mechanics*" (emphasis mine). Thus wrote Don Howard in his somewhat provocatively-named article "Who Invented the 'Copenhagen Interpretation'? A Study in Mythology" published in *Philosophy of Science 71* in 2004. Howard went on to write "Whatever Heisenberg's motivation, his invention of a unitary Copenhagen view on interpretation, at the center of which was his own, distinctively subjectivist view of the role of the observer, quickly found an audience." We might also add that it quickly became orthodoxy by setting up a filter through which Bohr's views were reinterpreted. Quantum physics is weird enough on any interpretation, but the really freakish ideas we saw earlier may be going too far even for quantum theory! Schrodinger thought so, as evidenced by his famous cat.

One critic compared Bohr's alleged position to that of Hinduism, presumably referring to the Hindu philosophy which understands the material world (or at least what we normally call the material world) as a mere illusion. In the same vein, Marcus Chown wrote, in the *New Scientist* article from which we quoted earlier "The pioneers of quantum theory, Niels Bohr and Werner Heisenberg, famously maintained that there was no real world out there and that quantum properties were brought into existence by the very act of measurement. Taken to its extreme view, the Bohr-Heisenberg view implies that the universe did not exist until we came on the scene to observe it." Notice how Chown assumes that Bohr and Heisenberg were united in their interpretation and that the actual subjectivism of the latter dates right back to Bohr and is by that fact soundly orthodox. But how can we square Chown's statement with Bohr's very own words that "It is certainly not possible for the observer to influence the events which may appear under the conditions he has arranged" (*Essays 1932–1957 on Atomic Physics and Human Knowledge*, p. 51), with its clear implication of the existence of objective events—and therefore of an objective realm—over which human perception has no control?

The answer probably lies in what historians of science have recently concluded about Bohr's philosophical presuppositions.

Tracing his intellectual development, it seems that the person whose thought exercised the greatest influence upon him was not a Hindu yogi but a much more likely candidate for being the intellectual father of a continental scientist; the German philosopher Immanuel Kant (1724–1804). Now, nobody is saying that, one fine evening, Bohr lit up his famous pipe, settled back in an armchair with a copy of the *Critique of Pure Reason* and later emerged with a fully worked out philosophy of quantum reality. We do not even know whether Bohr ever made Kant's opus a subject of specific study. In fact, the philosophers and historians of science who suspect a Kantian influence on Bohr also deem it unlikely that he was, so to speak, a conscious Kantian in the sense of one who deliberately and consistently subscribed to Kant's metaphysics. It is more likely that the philosopher's chief influence upon him was by and large indirect—via the media of several neo-Kantian thinkers—and that this influence bore fruit not so much in determining Bohr's belief system per se as in providing an intellectual framework within which he assessed certain issues raised by his research into the quantum realm. Nevertheless, this in no way diminishes its importance in guiding Bohr's thought.

According to Bryan Register, Bohr's principal contact with philosophy was through family friend and philosophy professor Harald Hoffding, whose two-volume work *History of Modern Philosophy* dedicates nearly 100 pages of Volume 2 to Kant. Hoffding clearly held Kant in considerable esteem and was influenced by his ideas, although he did not subscribe to the Kantian position himself. Nevertheless, it seems a good bet that any student of Professor Hoffding, or anyone else who may have engaged him in philosophical discussion, would have been given a fair dose of Kantian thought on which to ponder!

Therefore, in order to form a better appreciation of what Bohr most probably thought, and to avoid the extreme views that some of his professed supporters thought that he believed, it will be worthwhile to take a quick look at Kant's main ideas as well as briefly assessing his position in the development of European philosophy.

Western philosophy in the century or two prior to Kant could essentially be divided into two broad streams; the rationalist school (principally in continental Europe and most prevalent during the 1600s) and the empiricist school which flourished in Britain and principally during the following century. Both believed that philosophy was capable of arriving at a true picture of what the world was like, but they differed in their methods. The rationalists sought intuitive truths from which the system of knowledge could be derived by logical reasoning. For them, mathematics was the paradigm and they attempted to construct what might be described as a "mathematical" philosophy—not in the sense of one expressed in symbols, but one that used a similarly strict form of reasoning from indubitable propositions. By contrast, the empiricists took observational science as their model and believed that the way to attain knowledge of the world is ultimately based upon experience. Our reasoning is to be based, not upon self-evident "theorems" but upon the application of natural law; the knowledge of which is itself derived from our observations of the world.

Kant, although a continental philosopher, was closer to the empiricists, but he differed from both rationalists and empiricists alike insofar as he denied that either experience or pure reason could bring us knowledge of what the world is ultimately like. He divided the world into noumena (literally, "things thought of" but better described as "things in themselves" or "things as they really are") and phenomena or things as we experience them to be. All we can ever know are phenomena; things as we perceive them to be. Kant argued that we experience the world through what might be termed a system of mental filters and it is these "filters" or categories that impose upon the world the various features by which we describe it. Features such as space and time, causation and so forth do not, Kant argued, belong to the world as it really is, but only to the world as we experience it. It is as if we see the world through tinted spectacles, although the "spectacles" are worn by all our senses, not just that of sight, and they are inevitable aspects of our sensory perception and of our very thought processes themselves.

Nevertheless, it would not be correct to interpret Kant as simply saying that the realm of phenomena is merely subjective. This is demonstrated by his concept of space. Although Kant relegated space to the phenomenal rather than to the noumenal realm, he nevertheless argued in favor of Newton's thesis of absolute space as distinct from the relativity of G. W. Leibniz.

The Newton/Leibniz debate was quite a hot issue at the time and Kant came down on Newton's side by arguing that the absolute nature of space is revealed by such simple observations as the inability of left shoes to fit right feet or right-handed gloves to slip readily onto left hands. This, he claimed, proved that space has its own fixed properties in the manner that Newton held. These arguments do not affect relativity in Einstein's sense, as such features as left-handed and right-handed are given meaning within the same frame of reference, but Kant's argument did pose a difficulty for the more extreme relativity of Leibniz, which essentially denied that space had any objective existence at all. The Newton/Leibniz debate need not concern us here, but it is clear from Kant's contribution that he saw no inconsistency between his understanding of space as being both phenomenal and absolute or between his assertion that space is phenomenal and his denial of Leibniz's subjectivism. It might be beneficial to keep this in mind when considering some of Bohr's statements that might appear at first sight to be more subjectivist than he may have intended.

Kant was an agnostic, in the philosophical if not in the religious sense. Although popularly applied to the belief that the existence or non-existence of God cannot be known, in the philosophical sense, agnosticism simply means the doctrine that there are limits to knowledge and that some things (not necessarily God) lie beyond those limits. For Kant, what lies beyond the limits of knowledge is the real nature of the noumena. Kant had no doubt about the existence of the noumena per se, but it is probably no exaggeration to interpret his understanding of the noumenal realm as being a sort of permanently sealed box, the hidden contents of which remain forever unknown, and whose presence is discernible only through the filter of the categories.

Now, some philosophers (being philosophers!) have found an inconsistency here. Thus, in the very act of saying that something lies beyond the boundaries of knowledge, we are in effect saying that we already know one thing about it, namely that it lies beyond the boundaries of knowledge! Therefore, they argue, agnosticism is an impossible position in so far as its very statement involves one in a self-contradiction: viz. the implied assertion that one knows something about that which has just been declared intrinsically unknowable! The present writer sees merit in this argument, but now is neither the time nor the place to take it further. However, it will be worth remembering it in connection with what Kant and Bohr both said regarding the things which they claimed were unknowable. The unknowable nature of these must be less than absolute if this self-contradiction is to be avoided.

Bohr appears to have approached the subjects of quantum physics in at least a quasi-Kantian manner, though possibly not even being aware of his Kantianism. Quantum objects existed, but like Kant's noumena, their true nature lay outside the sphere of human knowledge or, with the above thoughts still in mind, at least outside the realm of direct empirical knowledge. Just as, for Kant, perception of the world at large imposed certain features upon the world as we *experience* it, so for Bohr our observation and measurement of the quantum realm impose certain structures upon *it*. It is with these structures that quantum physics deals. The noumenal atoms and sub-atomic particles are real and do not depend for their existence on their being observed by us, but their "phenomenal" features with which quantum physics deals are imposed upon our understanding of the quantum world by and through the process of our observation and measurement. Yet, just as we saw that Kant's concept of space was both phenomenal and absolute, even the immediate (phenomenal) subjects of quantum physics may not be as purely subjective as many interpreters of Bohr assert.

This is in line with Heinz Pagels' interpretation of Bohr's position in the following passage;

> We can imagine that quantum reality is like a sealed box out of which we receive messages. We can ask questions about the content of the box but never actually see what is inside of it. We have found a theory – the quantum theory – of the messages, and it is consistent. But there is no way to visualize the contents of the box. The best attitude one can take is to become a 'fair witness' – just describe what is actually observed without projecting fantasies on it. (*The Cosmic Code*, pp. 187–188).

Notice here that the messages coming from the black box are not spoken about as being simply in the mind of the observer. Without mentioning Kant, Pagels is nevertheless giving a very Kantian-looking interpretation of Bohr's approach to this subject.

It is because quantum physics deals with the phenomenal aspect of quantum objects that these can be considered as both particles and as waves, although not as both simultaneously. According to Bohr, the two sets of phenomena (pertaining to wave on the one hand and particle on the other) are complementary. A good analogy of this concept is the picture of a vase and profile of two human faces used by Gestalt psychologists. In this picture, the eye either discerns a vase or two faces in profile, but cannot discern both at the same time. Yet neither is an actual representation of what is really present. The reality in this instance is a sheet of black and white paper. Yet we cannot directly observe this as-it-is-in-itself and study its true properties (its texture, for instance) by simply looking at the image. We can only see it in terms of the picture; a picture which presents to our sight as two entirely different things—either two human profiles or a vase.

### Project 6: The "Gestalt" Analogy

What do you see when you look at this image? The profile of two human heads in silhouette? A single white vase? One and then the other? (Fig. 4.14)

FIGURE 4.14 What does the eye see? An image of a vase? Or two human heads in profile? (John Smithson)

> At times the image clearly shows two silhouettes of human faces. At other times, an "obvious" white vase is seen. Each image is a product of the brain's tendency to read some pattern into what it perceives and, to this degree, each image is dependent upon our observation. We could say that neither image exists unless it is being perceived, yet the images are not purely subjective or "imaginary" like, for instance, the snakes supposedly seen by a drunkard in a fit of delirium tremens. Something exists "out there"; something that, when it is being observed, is perceived either as the image of a white vase or as two silhouette profiles. But what is "really" present is simply a patch of black and white paper. This exists whether anybody is perceiving it or not.
>
> Although not perfect, this is a good analogy of quantum complementary and the relationship between the quantum phenomena observed and what "really exists" at quantum levels, according to the interpretation of Bohr presented here.

As an analogue of quantum complementary, we could say that the vase represents one set of phenomena (particle, for instance) and the profiles represent the other (wave) whereas the black and white paper common to both represents the actual objective world—the realm of noumena, in Kantian terminology. Both perceived phenomena are real in the sense that each is a true set of phenomena, but never shall the twain meet. Which set of phenomena is actually measured depends upon the nature of the experiment performed for that measurement.

We may read of a quantum physicist saying something like "phenomena are only real if they are observed". This, at first sight sounds like strict subjectivism, but it really allows two meanings, only one of which implies that philosophical position. Most often, when we speak of phenomena, we mean something that is present in the world. We might speak about "the phenomenon of the aurora", "the phenomenon of bird migration", "the phenomenon of the economic trade cycle" and so on. If we say that phenomena, in this sense of the term, only exists when observed, we can't avoid the radical subjectivism that Chown (incorrectly, we believe) sees as the ultimate result of Bohr's position.

However, if the statement "phenomena are only real if they are observed" is uttered by a Kantian, it takes on a whole new meaning. No radical subjectivism is implied and, indeed, the statement is essentially tautological. All it means is that "the world-as-we-observe-it is only real if it is being observed"; not a very profound statement by any measure! It is very important, when trying to come to grips with statements of this type, to understand where the person making the statement is coming from. In particular, any statements by Bohr that on face value appear to deny the existence of "phenomena" (in the usual non-Kantian sense) should be tested by giving them a Kantian flavor before too quickly assuming a subjectivist conclusion.

This Kantian interpretation of Bohr is summed up in his own words concerning the role of physics. He wrote that "It is wrong to think that the task of physics is to find out how Nature is. Physics concerns what we can say about Nature". Re-wording this using the technical terms employed by Kant, we get, "It is wrong to think that the task of physics is to find out about noumena. Physics can only concern itself with phenomena". I think that Bohr would have been entirely happy with that re-wording of his statement.

Bohr was never in danger of being scratched by Schrodinger's cat, nor would he have been too perturbed about Wigner's friend knowing whether a point of light appeared on the screen while, outside the sealed chamber, Wigner remained in ignorance of the experiment's result. For Bohr, the cat was either alive or dead prior to anyone opening the box and the pinpoint of light either did or did not occur irrespective of whether Wigner, or even his friend, saw it. Both the cat and the friend belonged to the macroscopic world and formed part of the phenomena (in the technical Kantian sense) of that world, whereas the collection of atoms that had a 50 % chance of emitting a radioactive particle, and that particle itself, belonged to the sub-microscopic quantum realm and constituted part of the distinctive set of phenomena (once again, in the technical sense of that word) that constitute "what we can say about [the quantum realm of] Nature". Once again, the vase and profiles of Gestalt psychology come to mind.

By the way, we might recall at this point what was said in an earlier chapter about the inaccuracy of the oft-quoted statement that, according to atomic theory, walking on a log of wood is like walking on a swarm of bees. The mistake in this can now be compared with an attempt to see both the vase and the profiles at the same time and to insist that only one of these is the true way of understanding the situation.

# An Intriguing Alternative Interpretation

Although the probability-wave interpretation of Schrodinger's wave function might be considered the orthodox or main-stream one in quantum theory, it is not the only one. Another, which actually seems to be gaining popularity in recent times, is the intriguing hypothesis put forward by Hugh Everett in 1957. Everett differed from the approach of the Copenhagen team in so far as he extended the wave function beyond the strictly quantum realm to also include the observer. The wave function itself, with all of its encoded possibilities, is the ultimate reality and the observer participates in each of them, even though he/she is only aware of one at each moment. The upshot of this is the implication that each observer has many histories, in many alternate worlds all branching out from each possible alternative. The old Chinese tale of the garden of branching paths is re-told in quantum language. Thus, if we are confronted at some particular moment by two mutually exclusive choices X and Y, although in this world we may choose X, in an alternate world splitting off from this world at the point of choice, we chose the alternative, Y. This idea was made the theme of the movie *Sliding Doors* where a woman's life split into two entirely divergent histories depending upon whether she reached a sliding door before it closed or whether she was in time to squeeze through. If Everett is right, Schrodinger's cat is alive in one branch of the universe and dead in another and Wigner's friend sees the tell-tale flash of light in one universe whereas a counterpart Wigner's friend—one that in one sense is and in another sense is not the same as the friend in our branch of the universe—sees no flash.

This interpretation was not well received at the time (not in *this* universe anyway). Everett received a less-than-encouraging reception from Bohr and the Copenhagen team and his paper was pretty much ignored by most other physicists. An exception was John Wheeler, who lent support to Everett and was initially favorably disposed toward the many-worlds interpretation, albeit not without some reservations.

Nevertheless, Everett's interpretation has gained somewhat in popularity in more recent years and is the preferred option for Massachusetts Institute of Technology cosmologist Max Tegmark, known for his remarkable work providing the mathematical tools for analysis of the CMB images that have become available in recent years. Outside of the science community, the idea of a multiverse has been popularized by authors of science fiction tales and movies. We have to admit, it does make for the basis of a good yarn, but is it more than this? Could this interpretation actually be true? It seemingly does have one advantage in so far as it gets rid of the primary role of probability and re-introduces a more deterministic model, albeit within a framework of ever-multiplying branches of the universe. It seems that if we wish to expunge the universe of weirdness in one respect, we must always pay for this action by re-introducing it in another form elsewhere. But then, isn't that just the price we must pay for the privilege of living in such a weird and truly fascinating universe?

Be that as it may, the Everett interpretation nevertheless raises an awkward issue. As we noted in the previous chapter with respect to another form of multi-world hypothesis, Pagels criticized the popular corollary of Everett's interpretation, namely, that all macroscopic possibilities are fulfilled in some branch of the universe, by arguing that even if Everett is correct, we cannot say that the many other worlds are truly distinguishable from ours. As quoted in Chap. 3 of this book, Pagels opines that the different worlds (granted for the sake of argument that they exist at all) might be related to one another like the different configurations of molecules in a gas, where the gas viewed as a whole appears the same for all the different configurations. If the many worlds all appear the same and exist in the same space and time, how can they be said to differ? (And they must exist in the same space and time as the splits occur at specific moments of time and points of space. Therefore, the framework of time and space in effect remains constant even while the different branches evolve in other ways within that framework).

Significantly, Wheeler eventually abandoned this interpretation on the grounds that "It required too much metaphysical baggage to carry around".

## What If the Wave-Function Is Real?

As was remarked earlier, the physical wave interpretation put forward by de Broglie and later developed further by Bohm, did not attract a significant following amongst quantum physicists, despite its early success in the prediction of the wave properties of particles. In that respect, it was not unlike Everett's interpretation but, also reminiscent of the Everett theory, it is far from being forgotten and might even be in line for resurgence.

That, at least, is the opinion of Matthew Pusey and Terry Rudolph of the Imperial College London, Jonathan Barrett of Royal Holloway University, also of London and Lucien Hardy of the Perimeter Institute for Theoretical Physics in Waterloo, Ontario.

What Pusey and his London colleagues did was to imagine a hypothetical theory that completely describes a single quantum system (for example, a single atom) but—and this is the crux of the matter—without the presence of a real wave directing the action of the quantum particle. They then concocted a thought experiment involving the bringing together of two independent atoms, making a particular measurement on them and then comparing the results (based upon their hypothetical wave-less theory) with the known outcome according to standard quantum mechanics. Their result was at variance with accepted—and well-tested—quantum theory. The team therefore concluded that their hypothetical wave-less theory must be wrong and that only a model that included a true wave could bring about the required outcome.

As team member Terry Rudolph remarked "Under seemingly benign assumptions, you cannot escape the wave function being 'real'". Anthony Valentini of Clemson University in South

Carolina agrees. According to this physicist, "[the team's work] shows that the wave function cannot be a mere abstract mathematical device. It must be real – as real as the magnetic field in the space around a bar magnet".

A slightly different, albeit complementary, approach was taken by Hardy. His chief assumption was what he termed "ontic indifference" i.e. "that quantum transformations that do not affect a given pure state can be implemented in such a way that they do not affect the ontic states in the support of that state". ("Ontic" being defined as "a particular state of reality"). In his paper, he argues that "any model reproducing quantum theory in which the quantum state is not a real thing" violates ontic indifference.

If the London team and Hardy are correct, the weirdness of probability waves is replaced by the same weirdness that Bohr et al. discarded back in the early days of quantum theory. The waves of de Broglie are real. Just as real as any waves that we encounter in nature, but far stranger! As we remarked earlier, Schrodinger's mathematics necessitates that the waves exist in an abstract hyperspace consisting of many dimensions. Only in the simplest case—the instance of a single quantum particle—does the wave function exist in three dimensions. For two particles, six dimensions are required; for three particles, nine and the whole things just takes off from there! As Valentini correctly remarks, "We need to expand our imaginations, widen our view of what constitutes fundamental reality". Unfortunately but perhaps understandably, he does not go on to suggest how we might put this suggestion into practice!

If the wave function is a real and objective fact it presumably means either that the host of alternate possibilities encoded within it exist in a sort of suspended reality, to be made real only by the collapse of the wave function due to some event (such as an observer making a measurement) *or* that they are all equally real, even if only one is actually observed. The first alternative seems to be harking back to the "if it is not observed, it does not exist" scenario whereas the second sounds suspiciously like the interpretation of Everett, though maybe not following him so far as to include the observer within the wave function. Nevertheless, if the wave is real, perhaps the most consistent interpretation is the one that understands all of the encoded possibilities as also being real in the same sense. Or maybe completely new interpretations will be put forward?

In any event, it seems vital to take note of Pagels' insistence that the quantum world and the macroscopic world are qualitatively and not just quantitatively different from one another and that what is permissible in one is not necessarily permissible in the other. In the macroscopic world, cats are either dead or alive, sliding doors either shut in our face or stay open just long enough for us to slip through. And while electrons and positrons can pop into being from the quantum vacuum and instantaneously annihilate each other, we do not see mountains of matter and their corresponding anti-matter counterparts emerging from the vacuum (fortunately for us, we might add!). In the macroscopic world, the positions and momentum of spacecraft and planets can both be predicted with a high degree of precision—otherwise mankind could never have sent a spacecraft to Saturn and dropped a probe on its largest moon, Titan. The ability to compute the trajectories of probes like this surely demonstrates the power of classical physics in dealing with macroscopic phenomenon (ranging from dust motes to superclusters of galaxies) just as surely as the computer technology necessary for the success of these same probes, as well as that by which images of the distant worlds themselves are transmitted back to Earth, is an apt demonstration of the power of quantum theory. Perhaps, with this in mind, we might dare to suggest that the universe may indeed have its sliding doors, but only quantum entities are able to pass through them. In other words, perhaps Everett was correct about the universe dividing at each moment into branching worlds, but maybe this only happens on quantum scales and to quantum entities. At the macroscopic level, all of these "alternate worlds" blur into a single universe with just one macroscopic history. Could the atoms of our body exist in a multiverse of many worlds? Sounds bizarre, but as the nature of the universe unfolds, that alone is not sufficient reason for disbelief.

Whatever the truth of the matter, we can be pretty sure that the near future will see many more papers concerning the possible reality of the wave function. Once more quoting Valentini, "You've seen nothing yet. Mark my words, this will run and run".

## "Spooky Action-at-a-Distance"

This was the phrase that Einstein rather derisively used to describe an apparent consequence of the standard interpretation of quantum theory. In 1935, Einstein, together with his colleagues Boris Podolsky and Nathan Rosen authored a paper detailing a thought experiment presenting what has sometimes been called the EPR paradox. The word "paradox" is actually a misnomer, as no real contradiction is involved, but the issue raised in the paper is real enough. Like Schrodinger's original purpose in his "cat" thought experiment, the EPR was also concocted to expose a problem with quantum theory; not just with the standard interpretation of the Copenhagen team, but with the underlying theory itself. Einstein believed that quantum theory was incomplete, and the EPR thought experiment was designed to prove this.

The EPR argument is as follows: Assume that two particles (for simplicity let them simply be titled 1 and 2) are near each other, with their distance from a common point of reference given as $q1$ and $q2$ (millimeters, feet or whatever units we choose) respectively. The particles are assumed to be moving and their respective momenta are denoted $p1$ and $p2$. Now, Heisenberg's uncertainty principle says that we cannot simultaneously measure $p1$ and $q1$ or $p2$ and $q2$ exactly. However, it does allow the accurate simultaneous measure of the *sum* of the momenta of both particles and the distance separating the two. The two particles are then imagined to interact before flying away from one another, each to the other side of the planet. Common sense and ordinary reason both agree that the two particles are now so far apart that nothing that happens to one could conceivably have any effect on the other. The alternative would break a cardinal principle of classical physics, viz. the principle of local causality—briefly stated, the principle that distant objects cannot instantaneously influence local objects without some form of mediation. Yet, if the thought experiment is looked at more closely, common sense and classical physics seem to be in for a nasty shock.

We know from basic physics that the total momentum of the two particles is the same after their interaction as it was before. In other words, it has been conserved. Therefore, if we were to measure the momentum (p1) of particle 1, we could then find the momentum (p2) of particle 2—on the other side of the world—by subtracting this measurement from the known total momentum. Measuring the position of particle 1 should (according to Heisenberg's uncertainty principle) disturb the accuracy of the measurement of its momentum but—according to the principle of local causality—should have no effect on our deduction of the momentum of the distant particle 2. But that implies (or rather, necessitates it would seem!) that we now know exactly the values of both the momentum and the position of particle 2, in violation of Heisenberg's principle! By sticking with the assumption of locality causality, we appear to have violated the uncertainty principle. But conversely, if we stay true to Heisenberg and quantum physics (that is to say, if we maintain the uncertainty), it would appear that our measurement of particle 1 must somehow alter the position or the momentum of the remote particle 2. Einstein and his colleagues concluded from this that either quantum theory conceals some causality-violating "spooky action-at-a-distance" or the theory is not the last word and needs to be supplanted by a more robust theory that will permit the simultaneous precise measuring of both the positions and momenta of quantum objects.

The possibility of "spooky action-at-a-distance" with its implication that a quantum event in one place can somehow instantaneously influence another at any point in the universe, seems to suggest a sort of "communication" at speeds greater than that of light (at *infinite* velocity actually). No wonder this thought did not please Einstein.

The choices seemed clear for the EPR authors. Either abandon local causality and the speed of light limit or conclude that quantum physics is incomplete. Guess which alternative Einstein and his colleagues chose!

The EPR paradox (problem would be a better term) remained a point of debate until 1965, when John Bell at CERN, near Geneva, tackled the problem anew and even proposed a real and physical, non-thought experiment by which it could be tested.

First, however, let's consider yet another thought experiment (this one devised by Heinz Pagels) to set the scene, so to speak, for Bell's insights.

Imagine, Pagels invites us, that we are in possession of a very peculiar type of nail gun. Unlike the more conventional variety, this one shoots nails, two at a time and in opposite directions, one from each end. This might seem a trifle dangerous, but the gun has an added peculiarity in that it shoots the nails sidewise, not like arrows as we might expect. Clearly, this imaginary gun would not be of much use on a building site, but it works very well for the proposed thought experiment.

Further imagine that, although each nail in a pair shot off simultaneously will have the same orientation, successive pairs will have completely random orientations.

The pairs of flying nails are aimed at two metal sheets (which we will, very unoriginally, call A and B), each of which has a wide slot cut in it that is large enough for each of the flying nails to readily pass through. The metal sheets are placed in front of, albeit at some distance from, each end of the nail gun. Located at a safe distance, two observers—one for each sheet—record the events once the gun starts shooting.

The slots in the metal sheets act as polarizers, in other words, as devices which allow objects having specific orientations to pass through them whilst blocking other objects that are otherwise identical but have different orientations. During the course of the experiment, the polarizers can be adjusted (the metal sheets can be turned such that the slots orient in different directions). The gun is set to fire and each observer keeps note of how many nails pass through the slots (hits) and how many do not (misses).

For a start, the two slots are set at the same orientation. Because the nails in each pair being fired has the same orientation and because the orientation of each pair randomly changes, there will be a certain number of hits and a certain number of misses. However, because the slots initially share the same orientation, both will record an equal number of hits and an equal number of misses. In other words, the two sets of completely random sequences will be precisely correlated.

For the next phase of the experiment, A is rotated slightly such that the angle of its slot differs by a small angle from that of B, which remains fixed. The gun is set shooting once more and each observer records the results. This time, because the slots have differing orientations, some pairs will include a nail that passes through A while the other member of the pair fails to get through B while other pairs will include a nail that passes through B as the other is blocked at A. The two random sequences will be different. If B is taken as the standard (because its orientation was not altered), A's mismatches with B (both hits and misses) can be termed errors. Let's say that there were 40 shots in which there were 4 errors (i.e. 10 % of the sequence).

The experiment is then repeated, but this time A is left in a fixed position and B is rotated through the same small angle, albeit in the opposite direction. The number of mismatches in *B's* record should this time match that in the above record of A, namely, 10 % of the total. This is because the configuration is identical in each instance.

Next, suppose that A is again turned through the same angle as in its original readjustment. The total relative angle between the two slots is now twice as large as it was previously. The experiment is once more repeated.

At first thought, it might seem that the percentage of errors in A relative to B should now be 20 %; twice as large as it was previously. This is simply the sum of the results of the two former runs of the experiment. Nevertheless, Pagels points out that by shifting A we have lost the standard for B's record and by shifting B we have likewise lost A's standard. Therefore, occasionally a double error will occur. For instance, suppose that a pair of nails would have registered a hit at both A and B if the slots were perfectly aligned. Because A is shifted from its original orientation, the nail will now score a miss at that slot. But since the slot in B has also been shifted, it is possible that the nail misses this as well and also shows up as an error. Two hits have been converted into two misses which, however, cancel out one another and are not recorded as an error. Therefore, because of the impossibility of detecting such a double error, the error rate with the angle between the slots set at twice that of the original will necessarily be less than the simple sum of the error rates of each of the separate changes in the angle of orientation of the slots.

This relationship is known as *Bell's inequality.*

Although the above is only a thought experiment, if someone wishes to set up a real experiment with flying nails and slots in metal sheets, the result is bound to agree with Bell's inequality.

## Project 7: Bell's Inequality

Pagels' nail-gun experiment looks good as a thought experiment and would undoubtedly demonstrate Bell's inequality if properly set up in the real world, however not many building-supply stores stock nail guns that fire nails sidewise simultaneously from both ends. Is there a simpler (and safer?) way of showing the inequality in the macroscopic world?

Using pulses of light instead of nails seems possible. Even simpler would be light reflected from some surface pattern. Images take the place of flying nails.

Let's try the following:

Take a cardboard cylinder such as the inside of a roll of paper towels or toilet paper. From a sheet of cardboard, cut out two circles and then cut rather large slots across the centers of both. Make sure that the slots exactly correspond. Next, draw a line on each cardboard disk perpendicular to the slots (this will enable the angle of the slots to be more easily determined by using a protractor).

Now, place a paper cap over each end of the cardboard cylinder and mark this with two spots of ink, each opposite each other and just inside the circumference of the end of the cylinder, such that the two spots are separated by a distance about equal to the diameter of the cylinder. Make sure that the ink spots at each end of the cylinder are in perfect alignment (these are to act as the counterparts of the slits from where nails were fired by Pagels' imaginary nail-gun).

Fix the two disks in an upright position and make sure that there is a clear view through both slots. Place the cylinder between them, so that when the spots at both of its ends are horizontal, each pair will be visible through the slot in both disks. We may say that the image of the pair of marks is passing through the slots, just like the nails in Pagels' thought experiment.

Now turn the cylinder through a wide range of orientations, trying to be as random in your choices as possible. An observer looks through the slot of each disk and notes whether both spots are visible. Having both spots visible is equivalent to having one of Pagel's nails pass through the slot. If both are visible, call this "1". Anything else, mark as "0". Mark the scores against "R" for the right-hand disk and "L" for the left.

Go through the same steps as for Pagels' thought experiment. When the two slots are each horizontal, both scores should show the same number of ones and zeroes. When one slot is turned through a small angle, a number of errors or mismatches should occur and when each slot is turned in opposite directions such that the angle is doubled, the number of errors will be greater although, according to Bell's inequality, less than twice the number of errors recorded when only one of the slots is turned at an angle and the other kept horizontal.

The writer's results were as follows.

When only one slot was turn, the following was recorded;

L 0100100100011011010011100011011110101011
R 1100101101011010010011101111011110101010

As you can see, there are 7 out of 40 mismatches here.

The results when both slots were turned by the same angle albeit in opposite directions were;

L 0100001011111010100111010101010011001010
R 0101011111111010101111011111010101011111

Here, there are 11 out of 40 mismatches, not 14 as we might have expected. Bell's inequality held.

But the million-dollar question is "Does Bell's inequality continue to apply if the flying nails are replaced by individual quantum objects?"

Before answering this, two very benign assumptions are made regarding the experiment with the flying nails. First, the nails are objectively real (they exist out there in the world of space and time quite independently of their observers) and, secondly, whatever happens to a nail at A has no effect on the corresponding nail at B (i.e. "local causality" is assumed). Both of these assumptions are called benign as they are not likely to be challenged be anyone considering the flying-nail experiment. They seem just too obvious to arouse controversy! But if someone actually set up this experiment and found that Bell's inequality *was* violated, one of these assumptions would need to be called into question as something decidedly weird must be going on. Most likely, we might imagine, the nails in that situation would be exerting some odd sort of influence on one another, or maybe some pervading force or whatever was affecting their orientation. Either way, somehow the orientation of each pair of nails would seem to be correlating in a way that simply should not happen.

Now, back to the quantum version. Bell himself expected his inequality to hold there as well. He was strongly of the opinion that the position taken by Einstein was the only rational one and that "spooky action-at-a-distance" could no more occur between quantum particles as it could between flying nails. Like Einstein, Bell thought that the apparent prediction that such a thing was possible implied that quantum physics had an inherent problem. After all, something that makes such a ridiculous prediction cannot be correct, can it? Not that either Einstein or Bell thought that quantum physics was wrong in the sense of being totally unable to account for observed phenomena. There was enough evidence even in the mid 1930s that this was not the case (as Einstein himself had already demonstrated). The real issue was whether it was complete or whether there was some as-yet-undiscovered "underlying" theory that would incorporate, yet pass beyond, quantum physics and restore classical determinism to the micro-real in a way that would uphold local causality and get rid of this apparent nonsense of action at a distance.

Bell's actual experiment mirrored Pagels' thought experiment in so far as it involved firing objects through a polarizer, except that in this instance the objects were photons and the polarizers were filters transparent to photons of specific polarizations. These polarizers can be rotated in much the same manner as the slots in the flying-nail thought experiment. This experiment, or some close version of it, has now been conducted a number of times and the result has been consistently surprising. Quantum objects do violate Bell's inequality; this is seen as evidence that quantum physics is indeed complete. There is no underlying theory waiting to rescue classical determinism. Somehow, in spite of Einstein, in spite of Bell and contrary to common sense, there really does appear to be "spooky action-at-a-distance".

The continuing puzzle is in what does this spooky action consist? There has been any number of suggested answers, some of them every bit as spooky as the action itself. Not surprisingly, the psychics and the hippies got hold of this as proof of telepathy and clairvoyance. Equally unsurprisingly, the less cautious notions were strongly challenged by more orthodox scientists.

Pagels takes what might be considered a minimalist perspective. He draws attention to the fact that (referring to a quantum counterpart of the nail experiment) what we have at A and B are two random sequences and that "If we asked out of all the things in the universe which one, if altered in a random way, would remain unchanged, the answer is: a random sequence. ...But by comparing [the random sequences at A and at B] we can see that there has been a change due to moving the polarizers—the information is in the cross-correlation, not in the individual records". He goes on to say that although the experiments do "imply that one can instantaneously change the cross-correlation of two random sequences of events on other sides of the galaxy" this cross-correlation itself is not a local object and the information that it may contain cannot be used to violate the principle of local causality. Neither information nor energy is transmitted instantaneously across space and the supposed "spooky action" is replaced by a not-so-spooky change in cross-correlation.

Others are not so sure. Thus Susskind has advanced the view that elementary particles may be connected by tiny wormholes (like miniature double-ended black holes, more formally known as Einstein-Rosen bridges) and even suspects that somewhere in all of this might lie the key to that pot of gold at the foot of the physicist's rainbow—a quantum theory of gravity!

But it might involve even more than that! Because the photons in the Bell experiments are produced and interact in such a way that the quantum state of each cannot be described independently, they are said to be entangled. This question of quantum entanglement is a very active field in physics and is attracting interest from fields as far apart as cosmology, quantum computing and encryption. Experiments have demonstrated that large molecules and even tiny diamonds ("macroscopic" objects in a broad sense!) can be entangled. But it may not stop even there. Recently, J. Maldacena at the Institute for Advanced Study in Princeton, together with Susskind, suggested that black holes could be entangled, just like the quantum objects in the Bell experiments! This could in principle happen in two ways. First, a pair of black holes could be formed simultaneously and be automatically entangled (not sure how this could happen in practice though, without the two coalescing) or, alternatively, Hawking radiation generated by one black hole could be captured and collapsed into a second black hole, thereby entangling the pair (also difficult to imagine in practice, if not in principle). Maldacena and Susskind see these entangled black holes as the two ends of a wormhole connecting two regions of space. If the black holes became entangled and then separated by large distances, it would be theoretically possible for spooky action-at-a-distance to be transmitted through the resulting wormhole across vast cosmic distances faster than the speed of light.

An even more remarkable candidate for entanglement was proposed by Laura Mersini-Houghton. We remember her hypothesis that the mysterious cold spot in the CMB and the great void lying in its direction may be evidence of another universe budding off from our own. We may now mention something we avoided when first this topic was raised. What she actually said was that our universe and this other universe (assuming this model to be correct) are entangled. When the parting of the ways between the two universes took place, both were still in the inflationary phase and their interaction was effectively between two tiny regions of the quantum vacuum. As inflation pushed out this tiny region to literally cosmic proportions, the imprint of its original quantum entanglement remained.

But why should we imagine that such an entangled universe should exist?

Mersini-Houghton arrived at this conclusion from an attempt to avoid the usual anthropic conclusion of the multiple universe predictions of string theory. We will be taking a look at string theory in the next section, but for the present it will be enough to say that one of the problems with this theory is that it actually describes a colossal number of vacuum states, each of which seems to be equally capable of blossoming into a universe. The total number comes out at around $10^{500}$. The usual way out is to take the anthropic argument that only a very small number of these universes possess the right physical laws to allow the presence of life and that our presence here is a consequence of "our" universe being one of these rarities.

For each patch of vacuum to become a universe, it must inflate enormously, probably (as widely hypothesized) due to the repulsive effect of vacuum energy—a sort of repulsive gravity that can theoretically be shown to arise when the vacuum pressure has a very high negative value. Contrary to what is widely thought however, Mersini-Houghton and colleague Richard Holman of Carnegie Mellon University in Pittsburgh, find that the dynamic effect of matter and gravity would weed out most of these enormous numbers of vacuum states, leaving only our patch and its close neighbors to inflate into true universes. Because these few remaining universes would have interacted when they were still tiny patches of quantum vacuum, they would remain in an entangled state with a spooky influence remaining between them.

Although the need for anthropic considerations appears less vital here, their model does not on its own explain the fine-tuning of physical laws necessary for the existence of life in the universe. We may still wonder why the dynamic effect of matter and gravity just happened to be such a fortunate combination. Once again, the philosophical (and, indeed, theological) issues raised earlier are brought to mind, but as this matter has already been pursued, it will not be further taken up here.

As we saw in the earlier discussion, the Mersini-Houghton explanation of the cold spot predicts a similar void in the opposite hemisphere that does not, however, appear to be present. On the other hand, Mersini-Houghton sees the unexpected patterns that emerged from the Planck data, including the infamous axis of evil as lending support to the existence of entangled universes literally pushing ours out of shape to a small but measurable extent. Whether this hypothesis is confirmed or not, the very fact that the phenomenon of quantum entanglement has been raised for elementary particles on the one hand and entire universes on the other demonstrates just how large the subject now looms in modern physics.

## The Universe with Strings Attached

Back in the 1960s, some physicists worked out a form of string theory in the hope of explaining the strong nuclear force that holds together nucleons (protons and neutrons) in the nucleus of an atom in such a powerful grip as to keep moderate numbers of like-charged protons together in the face of their mutual electrostatic repulsion. Strings, in the newer formulations of particle physics, are hypothetical one-dimensional objects that replace the point-like particles of earlier particle-physics formulations. They are understood as being the fundamental entities from whose quantum states elementary particles are thought to arise.

As it happened, this radical early proposal came to nothing and the string model of the strong force was replaced by an alternative theory known as quantum chromodynamics about which more need not be said here. Nevertheless, string theory itself was not thrown away, as it came to be realized that some of the problems which made it an unsuitable candidate for its original project rather ironically placed it in a very good position as a possible solution to an even thornier issue—a quantum theory of gravity. In particular, string theory predicts the existence of a particle which has very similar properties to those required for the hypothesized quantum gravity particle; the graviton.

As it developed in the latter years of last century, string theory became the hope of many physicists in search of the (thus far) elusive theory of everything. It began to look like a promising candidate, as it not only provided something that looked tantalizingly like a graviton, but also seemed to account for all the particles of the standard model of particle physics as well as the forces which operate upon them. Simply put, these particles could be treated as manifestations of the vibrations of the tiny strings, each of which is far smaller than a proton, possibly as small as $10^{-33}$ cm, although some physicists think that they may be somewhat larger, (the question of what composes the strings is better left unasked!). A potential difficulty—in trying to picture the strings and their twitching, if not necessarily in the theory per se—is that these string movements would need to take place in a hyperspace of ten dimensions to account for all of the required phenomena. This difficulty may be overcome however, if the extra dimensions—those other than the ones defining our familiar world—are compactified to very small values. To take a familiar analogy, a length of very thin cotton thread looks almost like a two-dimensional line, although we know that it is in fact a three dimensional object. It is simply that its thickness is so small as to be almost negligible in our experience.

A more serious problem however is that string theory developed into five different varieties. That was not a good sign for something that was supposed to explain all of nature within the parameters of a single theory. Neither is the prediction of a staggeringly large range of possible universes (noted in the previous section) that string theory was found to predict.

The situation improved somewhat after 1995, at least with respect to the multiplicity of alternative string theories. That was the year that physicist Edward Witten demonstrated that the five apparently disparate theories were in reality simply different aspects of a single underlying theory, now generally known as M-theory. Although still quite vague and to a large extent undeveloped, M-theory has become one of the hottest topics in fundamental physics today (Fig. 4.15).

FIGURE 4.15 Edward Witten 1951 –

Central to M-theory is the concept of a brane (not brain). The word is derived from membrane and is used to characterize an object of a certain number of dimensions when considered from a higher dimensional viewpoint. It is a subspace of a larger total space, the latter known as the bulk. For instance, a point particle is a 0-brane, a string is a 1-brane and something like the surface of the ocean might be termed a 2-brane wrapping the Earth and propagating in the four-dimensional spacetime of the solar system environment. The screen of a computer might likewise be termed a 2-brane existing within the bulk of our familiar three-dimensional space.

Introducing this concept into string theory sees the extra dimensions of this theory as large, although material particles and forces are confined to a three-dimensional space called the 3-brane. Because of this confinement, they fail to experience the true size of all but the familiar three dimensions. In effect, our universe lies on the three-dimensional surface of something like a membrane existing in a bulk space of higher dimensions. Indeed, M-theory requires one more dimension than string theory had earlier demanded, but if a ten-dimensional universe is accepted, adding an eleventh hardly raises any difficult issues.

One interesting idea that has been expressed in connection with some versions of M-theory concerns the possibility that gravity, alone of the forces of nature, may not after all be totally confined to the 3-brane on (or in?) which we live. Perhaps gravity leaks out of the brane and into the hyperspace in which our brane floats. If that is true, this might explain why gravity is so exceptionally weak.

Now, when first told that gravity is an exceptionally weak force, the layperson is often taken by surprise, apt to remark "gravity is not weak! Yesterday I fell down the steps and hurt myself quite badly, and not long ago I dropped a hammer on my foot and nearly broke my toe. Is that not proof that gravity is anything but weak?"

Let's look a little more closely at this. I once remember reading a comical poem—just a few lines of doggerel—about a debate concerning whether a fence be placed around a dangerous viewing platform or whether it would be better to permanently station an ambulance beneath to attend to people as they fell over the edge. The poem had a moral of course (preventing an accident is preferable to cleaning up the unfortunate results) but it also inadvertently made a good point of physics with the lines,

> It isn't the falling that hurts them so much
> It's the shock down below when their stopping!

Of course it is, although this is seldom recognized. What hurts when we fall or when something is dropped on our foot is the sudden cessation of the fall, not the fall itself. But what is it that really stops the fall? Not gravity, but actually one of the other forces of nature—electromagnetism. The matter in our bodies, in the Earth beneath our feet and in the hammer falling on one of our feet is made up of atoms, each possessing a negatively charged electron shell surrounding a positive nucleus. And, because like charges repel, the reason why a block of matter puts an end to our fall, to the fall of a hammer or to the fall of Newton's apple is ultimately down to electromagnetic repulsion. In short, we stop so suddenly that we risk injury because electromagnetism is so much stronger than gravity, not because gravity itself is a strong force. It is not the falling, under the influence of gravity, that hurts, but the shock when we are stopped ... or when *we* stop a falling hammer!

To appreciate the comparison between the different forces, if we let the strong atomic force be designated as "1", the electromagnetic force is 1/137, the weak atomic force $10^{-6}$ and gravity, an amazing $10^{-39}$. It is this tremendous difference in strength that has always been one of the problems with gravity. Maybe, opine supporters of some varieties of M-theory, it is really not so weak after all; just very diffused through a multitude of dimensions.

Although M-theory still has a long way to go, some physicists have waxed very enthusiastic about it, claiming that it may well be (or at least might become when fully worked out) the long sought after theory of everything. Others, however, are not so sanguine. One less-than-convinced physicist remarked that, given the strangeness of M-theory and string theory in general, he wonders if physicists 100 years from now will look back and marvel at the insights of their predecessors of the late twentieth/early twenty-first centuries or, will they just ask, "What were these guys smoking?"

Then, during a conference where M-theory was being discussed, another skeptical physicist was heard to murmur the following rhyming word of caution to his colleagues,

> Look to yourselves
> That you be not smitten,
> The book is not yet finished
> The last word isn't Witten!

Whether the last word is or is not "Witten" or more likely some further development of the theory first proposed by Witten, is yet to be determined.

# The "Big Splat"

One interesting development is the application of M-theory to the origin of the contemporary universe itself. Cosmologists Neil Turok and Paul Steinhardt put forward a model of a five-dimensional spacetime in which the four special dimensions are bounded by two three-dimensional walls, that is to say, by two 3-branes. One of these makes up the space in which we live. Between these two, there is a third 3-brane which, in effect, is loose and (very) occasionally hits one of the boundary 3-branes. A little less than 14 billion years ago, it crashed into our brane, releasing a great deal of heat through the effects of the collision. This burst of heat was (you guessed it!) the Big Bang, although a somewhat different Big Bang from the one usually hypothesized. For one thing, it did not begin in a singularity or some ultra-dense state approaching it. That is good news. But it also did not experience an inflationary era and, indeed, appears to rule out this possibility. That is not-so-good news. This cosmological model has officially become known as the ekpyrotic universe model and, more popularly, as the *Big Splat*. It is regarded as a cyclic model as the loose brane is thought to periodically impact the 3-brane that currently houses our universe, although the interval between collisions is long by comparison with the age of the visible universe.

There are some attractive features of this model, principally its avoidance of that troublesome initial singularity, however its inconsistency with the existence of an inflationary era presents a problem. Even though, as we remarked in the first chapter of this book, some physicists argue that inflation has sometimes been made to carry loads that it cannot properly bear, the theory itself still has a lot going for it.

Recently, it may have grown even stronger.

Astronomers have for several years believed that the smoking gun of inflation (if one exists) lies concealed in the CMB. As we saw in the first chapter, patterns in the CMB show clear evidence of structures that formed very early in the history of the universe and acted as templates for the chains of galaxies and large voids that populate the visible universe today. The discovery of these features was a remarkable advance in our understanding of the early universe, but what astronomers sought even more diligently was the pattern of a kind of wave that is predicted to have arisen during the inflationary era. These are the gravitational waves predicted by general relativity. In no way to be confused with the gravity waves or buoyancy waves that give rise to thin lines of cloud and bands in the airglow of Earth's atmosphere, gravitational waves are waves in the space-time continuum itself. They are believed to be produced in abundance by violent events such as the collapse of a star into a black hole, but they are very difficult to detect and have not directly been observed. Nevertheless, astronomers hardly doubted the existence of such things. For a start, their presence was firmly predicted by General Relativity and this has proven to be such a powerful theory in other respects that, if it says that gravitational waves exist, then gravitational waves exist.

Moreover, there is indirect evidence for their presence. Stellar objects having strong gravitational fields and orbiting a close common center of gravity are predicted to radiate gravitational waves. Two such objects are the binary pulsar (a pulsar and another neutron star in orbit around a common center) known as PSR B1913 +16, discovered in 1974, and the more recently (2003) discovered binary pulsar PSR J 0737–3039. Although current technology cannot directly detect them, both these systems should be radiating gravitational waves and, because these waves will be radiating away the energy of the binary systems, their orbits should be steadily decaying at a predictable rate. Indeed, both binaries are losing energy at just the amount predicted if gravitational waves are being generated. This is very good evidence for the reality of these waves.

Still, it would be good to observe gravitational waves more directly and their detection as ripples in the CMB would be one way of doing that. Such detection would strengthen General Relativity even further but even more importantly, they would supply the long-sought direct evidence of inflation and their continuing study should further refine the parameters of inflationary theory.

The ripples caused by these waves should be displayed in the polarization of the CMB. But the sought-after polarization has a different signature to the one that has been known for several years. The latter is known as E-mode polarization. Polarization produced by gravitational waves is known as B-mode, but there is a slight complication insofar as E-modes may actually be converted to B-modes through the gravitational lensing of galaxy clusters interspersed between Earth and the CMB. Therefore, even without inflation, a percentage of the E-modes will be picked up on Earth as B-modes. Fortunately however, this secondary B-mode polarization is distinguishable from what we might call the primordial B-mode arising directly from gravitational waves during the inflationary era.

The first variety of B-mode polarization was discovered by the South Polar Telescope team and officially announced in July 2013. This was an important discovery of course, but it said nothing directly about the existence of the inflationary era. Nevertheless, continued observation at the same observatory led to the big announcement, in March 2014, of the first detection of primordial B-modes in the CMB sky. If upheld, this will surely rate as one of the greatest astronomical discoveries in history. The patterns detected matched those predicted for primordial B-modes, although they were somewhat easier to detect (stronger) than anticipated. No doubt the further analysis of this fact, assuming that it will be confirmed, will uncover further information about the nature of inflation and continuing study will assist in refining theories as to the nature of this process. As remarkable as it seems, through observing these gravitational waves, the SPT team may well be observing the Big Bang itself (Fig. 4.16).

If subsequent observational evidence confirms it, astronomers see this discovery as the finding of the smoking gun of

FIGURE 4.16 The South Pole Telescope (Report of the US Antarctic Program Blue Ribbon Panel. Credit: NASA)

inflation. But we may also note that this same gun has just shot the ekpyrotic universe model. If there really was an inflationary epoch, the model that excludes it must be incorrect. If the wound is not to be mortal, somehow the ekpyrotic model must convincingly come to incorporate inflation. At present, this does not appear likely, but who can guess how theories will develop in the future?

Assuming that the ekpyrotic model cannot be resuscitated, does that imply that the universe is a no-braner—at least in terms of M-theory branes? Not at all. The ekpyrotic model requires M-theory to be valid, but is not in itself a necessary prediction of this theory. If the Big Bang was not also a Big Splat, the hypothetical floating brane that periodically bounces off our own 3-brane presumably does not exist, but that does not necessarily mean that M-theory itself is incorrect. Maybe further development of the theory will produce another model of the Big Bang, differing from the ekpyrotic model and maybe not just incorporating inflation but actually explaining how and why this process occurred.

# 5. Observations and Ideas from the Left Field

The universe is a weird place, at least by the standards of our everyday experience. Yet, there is a general consensus that the picture painted by modern science, both observational and theoretical, is more or less true to reality. Strange and counter-intuitive though much of the theory is, within the range of size, velocity and energy that defines what we might call our familiar world, it nevertheless appears to explain that world of everyday experience quite satisfactorily. Things only truly get weird when we pass beyond this and deal with dimensions, velocities and energy levels that pass far beyond our normal experience. Then we encounter the apparent weirdness of relativity, quantum physics and so forth.

That is not to say that everything is settled. Scientists accept as an act of faith that the universe really is describable in terms of a single theory that is yet to be discovered. This theory, when discovered (hopefully when and not if) will combine the two great physical theories of relativity and quantum mechanics; one describing the universe on its grandest scales and the other on its most minute. At present, it is almost as if there is a relativity universe and a quantum universe, but our everyday experience tells us that these universes are just different aspects of the one and that somehow these two theories must also meet.

Other issues also remain unsolved. Is string theory on the right track? Is M-theory? Inflation appears to be correct, but can we really be sure of this? And, if it is indeed correct, which version of the theory is the true one? Or has that one not been found as yet? Does space have a non-trivial topology? What is dark matter? And dark energy? These are all topics under study and awaiting answers.

Although nobody can be absolutely certain, the consensus is that the answers to these and other questions—many of which are still waiting even to be asked—will lie more or less within the boundaries of the present model of the universe. Almost certainly, there will be some alterations and readjustments, but it is fair to say that fewer and fewer cosmologists expect a total revolution in our thinking about the universe.

Nevertheless, there have always been some who have not agreed with current orthodoxy. This is a good thing. If nothing else, it keeps researchers alert to difficulties that may otherwise have slipped by their attention. Sometimes an alternative theory will be put forward that appears to explain accepted data as convincingly as the more generally accepted one or even has some apparent advantage over "accepted wisdom". At other times, observations are presented that, if upheld by further research, would seriously challenge accepted beliefs.

We have already looked at some instances of each. The Steady State theory is a prime example of an alternative theory to the one that eventually became "orthodoxy". For a time, the Steady State seemed to have more going for it and attracted more supporters than the Big Bang; until further data altered the balance in the latter's favor. An example of the second was the controversy over quasar redshifts.

In this chapter, some more examples of both are looked at. Some of these left field ideas and observations are now pretty much dead issues, although they remain important in so far as the challenges they offered played an important role in our evolving understanding of the universe. Others are still very much of relevance.

A good place to begin is with one of the subjects briefly discussed in the first chapter of this book; the controversy over discrepant redshifts. We looked at this with respect to quasars, but the issue actually was broader than this and if some of the data presented had been upheld by later observations would have given us a very different picture of the universe than the one we now possess. Whether that universe model would have been more or less weird than the one we have is not known.

# Odd Redshifts

Strange Companions: NGC 4319 and Markarian 205

Beyond the stars of the constellation of Draco lies a rather unremarkable looking barred spiral galaxy catalogued as NGC 4319. At least, the galaxy is quite unremarkable except for one thing. Very close to its edge there is a quasar catalogued as Markarian 205. Even that is not especially noteworthy. At least, not until the redshifts of each object is measured. Being so close together in the sky might lead us to think that the two are truly close together in space, but it is here that the remarkable aspect raises its head, for while the redshift of NGC 4319 is moderate, that of Markarian 205 is large. If each redshift is interpreted as cosmological (that is to say, solely due to the expansion of the universe) the distance of the galaxy comes out at around 77 million light years whereas that of the quasar clocks in at one billion.

The chance that this is a simple co-incidental alignment between two objects of vastly differing distances looks, on the face of it, pretty remote. Too remote indeed for astronomer H. Arp and his colleagues. Neither did Arp's skepticism rest solely on the low probability of this alignment alone. By closely examining images of the galaxy/quasar pair, he claimed that a faint "bridge" connected the quasar to the outer edges of the galaxy and even suspected a slight distortion in the galaxy itself which almost, but not quite, aligned with the quasar. In Fig. 5.1, the brighter inner region of the galaxy does indeed appear slightly elongated, in a direction slightly angled to a line connecting the galactic nucleus and the quasar. This suggested to Arp and his supporters that the latter may have been ejected from the nucleus in the relatively recent past (in terms of the cosmological calendar of course) and that the continued rotation of the galaxy during the intervening period accounts for the present small misalignment of the distortion caused by the ejection event.

This sounds rather plausible but it does leave open some important questions. For one thing, what actually is the quasar? Is it a ball of gas, a compact cluster of stars or some sort of single-object megastar? How was it ejected? And why? Furthermore, what was it that happened in the nucleus of NGC 4319 causing

216  Weird Universe

FIGURE 5.1 The "odd couple"; NGC 4319 and Markarian 205 have vastly differing redshifts, despite their very close proximity in the sky (*Credit*: NASA/ESA & Hubble Heritage Team STScI/AURA)

the ejection? Whatever it was, the very fact of something as large as Markarian 205 shooting out through the spiral arms into intergalactic space must surely have resulted in major disruptions of the galaxy's structure. A slight elongation that almost needs to be seen through the eyes of faith and a weak luminous bridge just doesn't seem sufficient. In fact, we can now dispense with the luminous bridge. Further analysis of the images which Arp purported to show this feature strongly suggest that the bridge is nothing more substantial than an image artifact.

A more general issue is why a galaxy as normal as NGC 4319 should spit out a quasar in the first place. Or, to put it another way, why, if this galaxy could perform such a trick, does not every similar galaxy in the universe shoot out quasars left, right and center? Our own Milky Way is a barred spiral of similar type to NGC 4319. Where are its quasars? Or are they numbered amongst the quasars which we see scattered throughout the sky? But then, as already raised in Chap. 1 of this book, why don't we also see a goodly number of blue-shifted quasars that have been shot forth from large nearby galaxies such as those in Andromeda and Triangulum?

True, the NGC 4319/Markarian 205 pairing is not alone. The object initially listed as radio source 4C-1150 consists of a pair of quasars separated by just 5 s of arc on the sky, yet (assuming their redshifts to be cosmological) a factor of nearly five in real distance. And there are several other instances of quasars having high redshifts being found apparently near galaxies having low ones.

Nevertheless, the consensus of opinion now favors chance alignments. In the NGC 4319/Markarian 205 instance, observations of the quasar made with the Hubble Space Telescope have detected absorption lines in its spectrum corresponding to the halo of NGC 4319, indicating that the quasar lies beyond the galaxy. That does not prove that it is hundreds of millions of light years beyond of course, but in the absence of any strong evidence that the redshift of this object is not cosmological plus strong evidence that other quasars are at the distances which their redshifts indicate, it is a fair conclusion that chance alignment is the lesser of two evils.

Considering quasars in general, it is worthwhile remembering that, despite some curious alignments with galaxies having low redshifts, there are many more instances of association with galaxies having similar redshifts to the quasars themselves. As telescopes reach deeper and deeper into cosmic space and fainter and fainter objects are discerned, galactic companions of quasars have increasingly been found. Even the host galaxies of some quasars have been detected and there can be no doubt that these objects really represent a very energetic phase in the life of certain galactic nuclei. When the image of the bright nucleus (the quasar itself) of 3C 273 is blocked, the underlying galaxy is identifiable as a giant elliptical, not unlike the far closer M87 in the Virgo galaxy cluster (see Fig. 5.2). The only possible way to explain the

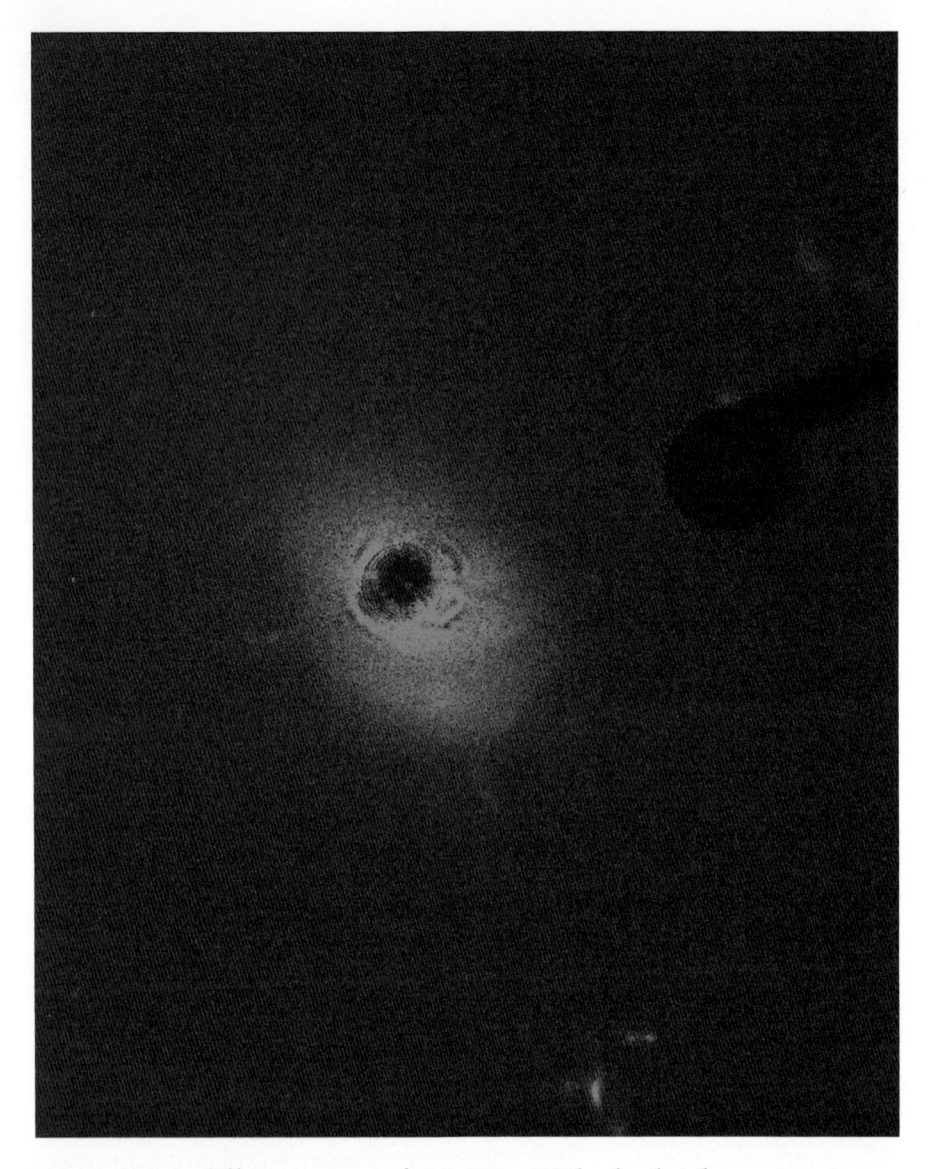

FIGURE 5.2 A different view of 3C 273. With the bright quasar image removed, the host galaxy—a giant elliptical—becomes visible (*Credit*: NASA & ESA Hubble Telescope)

existence of quasars ejected from nearby galaxies is to assume that there are two very different classes of object subsumed under the name of quasar. This possibility was actually put forward at one time and it must be said that there is some historical precedent for it.

Back in the days before astrophotography and spectroscopy, fuzzy patches and blobs of light in the sky were all lumped together under the name of nebulae and there was some debate as to whether these were relatively nearby cloud-like objects or distant, vast, systems of stars known then as "island universes". As it turned out, they are both. Some nebulae (true nebulae) are clouds of gas (like the grand one in Orion) or dust (for example, the wisps and dimly glowing patches within the Pleiades star cluster) within our galaxy. Yet others are dense clusters of stars relatively close to our region of space while others again proved to be true island universes or galaxies as we prefer to call them today. Perhaps, some people argued, quasars are not dissimilarly divided. Granted that some are the hyper-active nuclei of very remote galaxies, is it not possible that others are objects that, although superficially resembling the first kind in appearance, are nevertheless very different in their nature. Exactly what the objects of this second kind are is another issue, but they are apparently relatively small compared with galaxies and are somehow ejected in some unknown manner by galaxies that appear normal in every other respect.

The problem with this suggestion is that, apart from the discrepant redshifts themselves, there is not the slightest shred of evidence to support it and several reasons (as already discussed) to doubt its validity. Not surprisingly, it has never boasted a great following amongst astronomers.

## Discordant Tunes from the Quintet (and the Sextet)

Back in 1877, Edouard Stephan of Marseille Observatory discovered an unusual cluster of five nebulous patches in the constellation of Pegasus. What he had found was a previously unknown astronomical phenomenon, although its true nature would not be fully realized for a good many years. Stephan's Quintet, as it has long been called, is an example of a compact galaxy group, not only the first known but, in the century-plus since its discovery, also the most thoroughly studied.

The cluster is a picture of turmoil. Four of the quintet's members are obviously being disrupted by each others' gravitational attraction. Two have almost merged; their nuclei peering out like two eyes of a smiley face! The smile is one of several arcs that festoon the cluster. These are not Mandl arcs, but genuine features of the cluster's environment, believed to be shock waves, rather like sonic booms caused by one of the galaxies falling inward through the intergalactic medium at a velocity of over one million miles per hour!

But just outside this group of disrupted and colliding galaxies, lies the somewhat brighter fifth member of the quintet, serene by comparison with its neighbors.

Astronomers were in for a surprise when first they determined the redshifts of the members of this group. The four disturbed members presented no problem, as they all were found to have similar redshifts placing them somewhere between 210 and 240 million light years away. A sixth object somewhat removed from the compact cluster is at a similar distance and is probably related to the group. However, the brightest member of the group (NGC 7320) revealed a far smaller redshift than its companions. Interpreted as cosmological, its apparent velocity of recession placed it just 39 million light years, or thereabouts, from Earth, well short of the distance of the other group members. Is this galaxy simply a foreground object having no association with the compact cluster or is there something more going on here?

Champions of the non-cosmological interpretation of discrepant redshifts suspect the presence of exotic physics or something equally as weird here. Simple ejection, even from such a gravitational maelstrom as Stephan's quintet, does not appear capable of ejecting an entire galaxy. Especially not one as neat and undisturbed as we have here. Even if quasars—or objects that superficially resemble quasars—could be hurled out of galaxies into surrounding space, galaxies are a different matter.

No such weirdness is required however. Thanks to higher resolution images of this system, star-forming regions have been resolved in the low-redshift member, and their size and brightness correspond well with what would be expected for such objects at the distance indicated by that galaxy's redshift. In fact, looking at modern high-resolution images of the group, this galaxy is seen as standing out from those around it. It is indeed the odd member of the quintet and it comes as no great surprise to find that it is simply a foreground object, even though such an alignment is truly a remarkable coincidence.

## Project 8: Stephan's Quintet

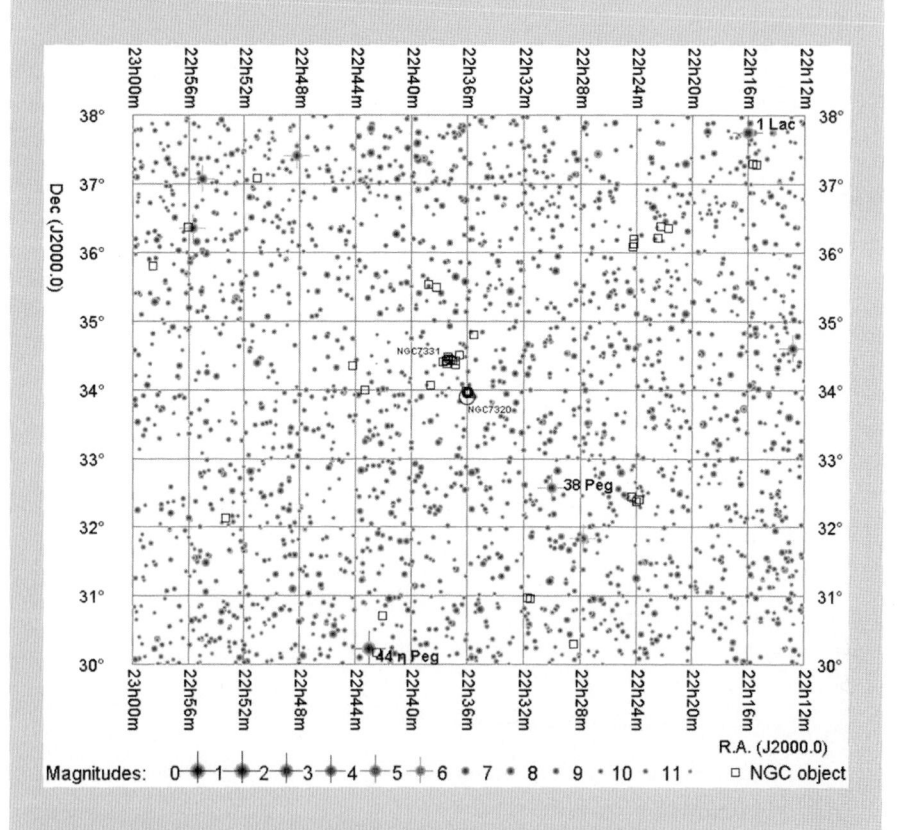

Stephan's Quintet is visible in moderate-sized telescopes from suitably dark sites. The interloper in the group, designated as NGC 7320, is around magnitude 13 and has been detected by experienced observers using telescopes of 10 in. (25 cm). The bona fide cluster members are fainter, estimated as between magnitude 14 and 14.6 but these have also been found, again by experienced observers under good skies, with telescopes of 12 in. (32 cm). Use the brighter NGC 7320 as a signpost for its fainter companions.

The above chart, plus the image of the cluster in Fig. 5.3, should help locate this relatively faint group of galaxies.

222  Weird Universe

FIGURE 5.3 Stephen's Quintet (*Credit*: NASA)
Finder chart courtesy of Graham Relf (www.grelf.net).

Stephan's quintet is not, however, alone in being a compact cluster of galaxies with an apparent member having a strongly discrepant redshift. An even more remarkable cluster was discovered in 1951 by Carl Seyfert whilst examining photographic plates originally secured at the Barnard Observatory of Vanderbilt University. Coincidentally, the brightest member of the group (a lenticular galaxy now catalogued as NGC 6027A) was actually discovered as far back as 1882 by none other than E. Stephan of Stephan's Quintet fame. He may even have spotted two of the other cluster members as well, as he refers to a couple of "stars" close to the lenticular, apparently in the location of brighter spots within the other group members.

Now known as Seyfert's Sextet, this cluster was, at the time of Seyfert's discovery, the most compact known and its members (all except one, as we shall see) are currently tearing each other apart as their component stars and interstellar material merge into what will in the future become a single elliptical galaxy.

Actually, although smaller instruments appear to show six individual objects, deeper images reveal two of these to be in reality parts of a single tortured galaxy. This object is elongated with a brighter portion sitting at the end of the elongation, masquerading as a separate galaxy in instruments too small to discern the fainter sections of the galaxy's elongated body.

Remarkably, apparently enmeshed in all this gravitational turmoil, lies a small and neatly defined face-on spiral galaxy.

Reminiscent of Stephan's cluster, it is this spiral that shows a vastly discrepant redshift. However, unlike the former instance, this one has a significantly greater redshift than its tormented neighbors. While the other members of the group sport redshifts indicating a common distance of around 190 million light years, the little spiral clocks in at a very large 877 million light years or thereabouts.

Once again, the coincidence seemed too great for some astronomers and those seeking evidence for discrepant redshifts of a non-cosmological nature were ready to see something peculiar here. Nevertheless, it must be said that, once again, the galaxy having the discrepant redshift stands out as being very different in appearance from the others in high-resolution images. It is very difficult to understand how it could retain its neat whirlpool shape in the sort of environment that exists within so compact a group of interacting galaxies. Moreover, if its redshift truly is cosmological, its intrinsic size is more representative of a spiral galaxy of this form. Reducing its distance to that of the bona fide members of this cluster would so expand the apparent dimensions of a normal spiral galaxy as to cover much of the remainder of the cluster. There is, in short, little doubt that the spiral is a remote background field galaxy that just happens to lie along the line of sight of the compact cluster. The coincidence might be weird, but the weirdness stops there. No remarkable ejection from other galaxies or unconventional physics is involved.

## Project 9: Seeking Seyfert's Sextet

This is a challenge for experienced deep sky observers blessed with good skies and good instruments. The cluster itself lies in the constellation of Serpens Caput, close to the boundary between this constellation and Hercules at the position (2000 equinox)

| | | |
|---|---|---|
| RA 15 h | 59 min | 11.9 s |
| +20° | 45 min | 31 s |

The brightest member of the cluster is NGC 6027A and, unlike the most prominent "member" of Stephan's Quintet, really *is* a bona fide part of the group! Unfortunately for visual observers, its magnitude is estimated as just 14.7. A 12-in. (32-cm) telescope might reach it and both this galaxy and two of its fellow group members have been seen by at least one experienced observer using a 17.5-in. (44-cm) telescope. Experienced visual observers of faint galaxies might like to test the minimum size of telescope capable of finding this galaxy and at least its two brighter companions. The faintest member of the group is around magnitude 17; a very difficult challenge for visual observation.

## The Strange Case of VV 172

A further instance of discrepant redshift, possibly the most remarkable of them all, is the chain of five faint galaxies (the first galaxy chain to be recognized) known quite unpretentiously as VV 172—and also as Hickson 55—located in the constellation of Draco. Discovered by B. A. Vorontsov-Velyaminov in 1959, the two brightest galaxies of the group had their redshifts first measured in 1960 by husband and wife astronomers G and M Burbidge and found to be approximately the same, confirming the suspicion that the chain is a real physical association lying some 700 million light years away. Eight years later, the redshifts of all five were

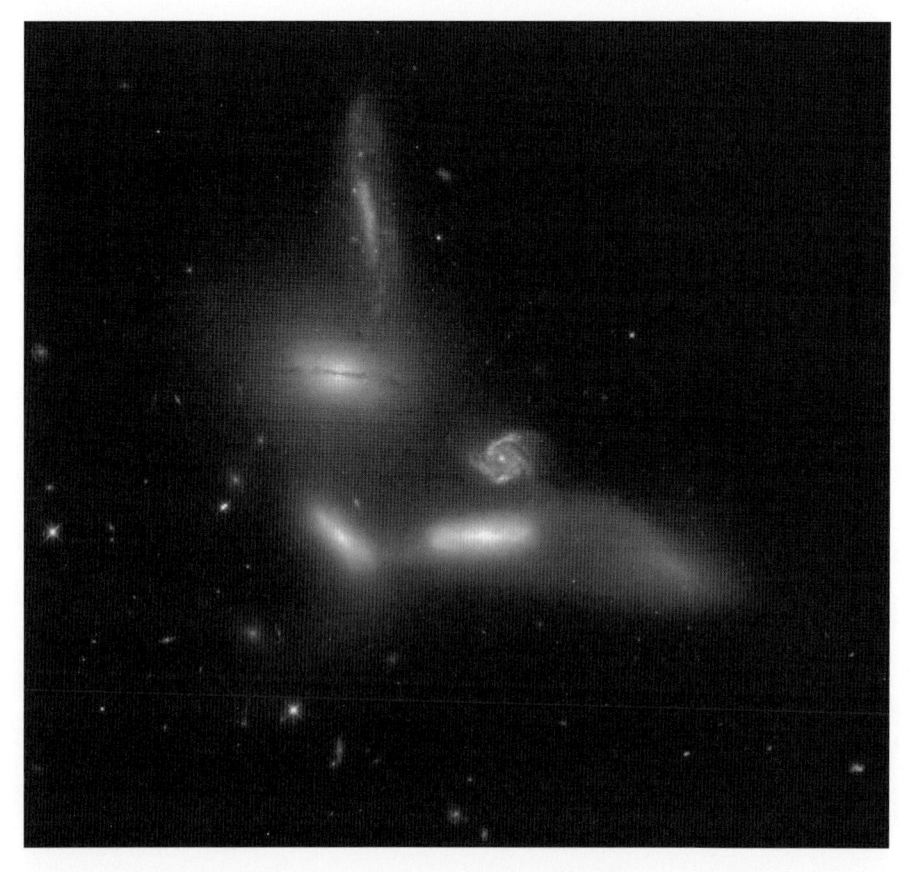

FIGURE 5.4 Seyfert's Sextet (*Credit*: NASA & ESA)

measured by W. Sargent who confirmed that four of the objects were indeed located at quite similar distances. However, the smallest and faintest member of the chain—the one located second from the top in Fig. 5.4—told a different story. Its redshift was found to be much higher than the others. If truly cosmological, this object is a whopping 1.5 billion light years from the Milky Way. Sargent remarked that "Clearly, something very odd is going on here!" But in what did this oddness consist? Sargent suggested three possibilities, viz. A chance coincidence, gravitational redshift or a true Doppler shift produced by the high velocity of this galaxy away from the rest of the chain. He quickly eliminated the gravitational possibility, as this would involve an unrealistic concentration of mass in the faintest member of the chain. The chance alignment

FIGURE 5.5 The galaxy chain VV 172 (aka, Hickson 55). The second galaxy from the top has a redshift greatly exceeding those of the four other objects (*Credit*: Ray Cash)

also appeared highly improbable, so maybe the errant member really is flying away from the rest at high speed, although why that should be so is unclear (Fig. 5.5).

Interestingly, Sargent also drew attention to a slight trend in the redshifts of the other chain members. Although very similar to one another, there is a slight increase along the chain with the two closest to the discrepant member (one on either side of it) having slightly larger redshifts than the remaining two. It has been suggested that this might indicate rotation of the chain, however such slight differences are hardly comparable to the large discrepancy exhibited by the faintest member of the group. Rotation is hard-pressed to explain that!

It has also been suggested that the errant galaxy might be a very faint background object magnified by the gravitational lensing of the two galaxies that appear to flank it. Objects of equal remoteness must cover the background field, but by and large remain too faint to be detected in normal telescopic images. It is doubtful, however, if there is enough intervening mass in the galaxy chain to account for significant gravitational lensing of one of these very distant galaxies.

Despite the chances stacked against it, the most likely explanation is probably the least exotic; the galaxy having the discrepant redshift is simply a very large, very remote, background object. Other similar field galaxies are located in the region of the chain and it seems that, improbable though it surely seems, one just happens to lie in direct line of sight with this chain of closer objects.

## Quantized Redshifts; An Even Bigger Headache for Cosmologists!

Discrepant redshifts would be bad enough and most cosmologists must have breathed a sigh of relief as other simpler and (dare it be said?) more orthodox explanations came to the fore. However, the problems that these discordant observations might have caused were as nothing compared with the apparent implications of a survey of galactic redshifts carried out in the late 1960s/early 1970s by astronomer William Tifft. In 1973, Tifft announced the result of his survey of over 200 redshift measurements of galaxies, many of which belonged to the great cluster of these objects found within the constellation of Coma Berenices. What he found was evidence for the clustering or banding of redshifts at certain values. This, he concluded, provided evidence that the redshift of galaxies "has properties inconsistent with a simple velocity and/or cosmic scale change interpretation". In other words, there seems to be something distinctly odd about the cosmic redshift and its usual interpretation in terms of an expanding universe. Whereas, assuming the usual interpretation, we should expect a steadily increasing value of redshifts with distance, what Tifft was finding seemed to hint that the redshifts increased in discrete jumps—quantum leaps, so to speak! Tifft basically noted this and, although drawing some implications from these results, did not offer any explanation for these strange findings.

The thought of quantized redshifts was less than enthusiastically embraced by the astronomical community. No doubt, this was in part due to reluctance to open the floodgates to whatever implications this phenomenon, if proven, might have but it was also true that the evidence presented fell far short of what would be required for such a revolutionary concept to be taken seriously. Moreover, as one skeptical astronomer remarked "Something is either quantized or it is not" and Tifft's results revealed, at best, what could be called a tendency toward quantization rather than strictly discrete quantum leaps in the data. Moreover, the sort of thing that Tifft appeared to be finding was truly difficult to explain in terms of what we know—or *think* that we know—about the universe.

Nevertheless, further work by several other astronomers, observing between 1989 and 1997, appeared to support Tifft's results. Thus, a 1989 study by M. Croasdale using a different sample of galaxies from that studied by Tifft also found evidence of quantization of redshifts. A study of bright spiral galaxies by B. Guthrie and W. Napier published the following year reported a "possible periodicity" in their results and, 2 years later these same astronomers found evidence of a similar periodicity in a sample of 89 galaxies. Other evidence of periodicity or quantization was given in studies by G. Paal and A. Holba et al. (1992), A. Holba (1994) and, most impressively, again by Napier and Guthrie in a study of 250 galaxies that they published in 1997. Concluding this latter study, these astronomers wrote that "the redshift distribution [of the galaxies studied] has been found to be strongly quantized in the galactocentric frame of reference".

However, and possibly of significance, not all of the suspected quantum leaps shared the same magnitude. Some studies suggested that the galactic redshifts took bigger leaps than others.

Observations and Ideas from the Left Field 229

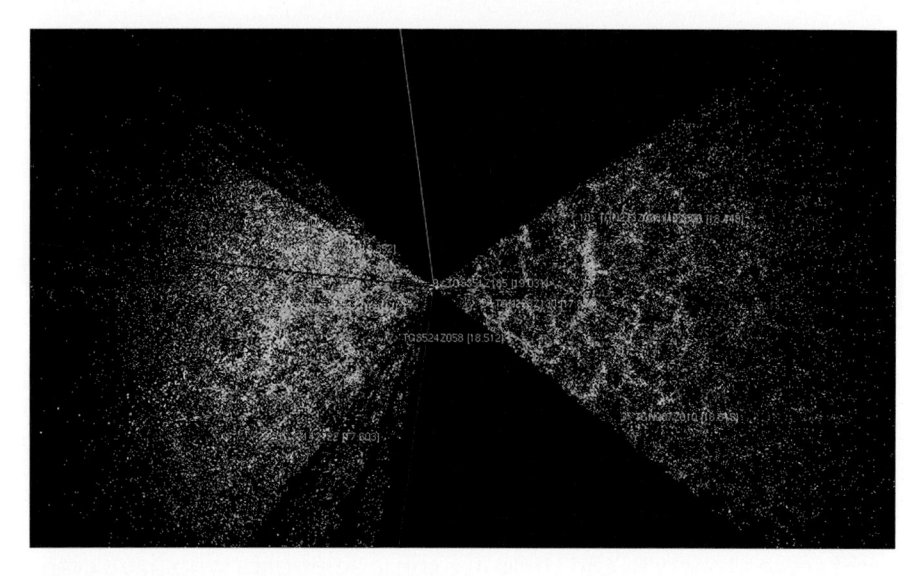

FIGURE 5.6 The 2dfgrs galaxy survey fails to show any real evidence of redshift patterns (*Credit*: Enoch Lau)

Since the 1990s, new observational technology has meant an explosion in the amount of galactic redshift data, and studies since 2000 have consistently seen a weakening of the evidence for quantized redshifts in the face of the large galaxy surveys that have now been conducted. If the phenomenon is real, this is not the way that things should have gone. A real pattern becomes clearer and more distinct as increasing data are accumulated. The contrary, as has happened here, implies that the pattern supposedly detected was not real and had more in common with the pseudo-patterns that seem to emerge from white noise than with something that is actually present (Fig. 5.6).

# Gravitational Weirdness

The enormous gap in strength between gravity and the other fundamental forces of nature has already been remarked upon. No generally accepted solution to this problem has been found. The suggestion that has already been raised (gravity extending beyond the brane in which our universe exists) is one possibility if M-theory is correct, although even then we must remember that this hypothesis is not an inevitable consequence of M-theory. It is just one possible way out of the problem that may be opened up if M-theory is proven to be correct.

But an even greater problem with gravity concerns the difficulty in finding a theory that unites it with the other forces of nature. This is not unrelated to the weakness issue. If all the forces of nature, including gravity, really are aspects of a single force that manifests at high energies, then the strength (or rather, lack) of gravity means that the energies at which it is united with the others must be very high indeed; not something that can be reached even with the LHC.

Furthermore, a unified field theory that includes gravity depends upon a satisfactory quantizing of gravity. This remains a big stumbling block. No theory of quantum gravity yet put forward seems to work, and there are some scientists who have come to doubt that quantum gravity is even possible at all. Hal Puthoff for instance, points to the increasingly involved mathematics required in the formulation of theories of quantum gravity. The theories seem to be getting more and more abstract and less and less intuitive while still (at this moment at least) falling short of a satisfactory solution to the issue. That gravity should be incapable of quantization would be very strange indeed, but in this weird universe, that in itself does not necessarily count as an argument against it! (The recent and, at the time of this writing unconfirmed, primordial B-mode polarization detection in the CMB does, however, appear to require the quantization of the gravitational waves that gave rise to these B-modes. The apparent B-mode detection has therefore been hailed by certain scientists as the only evidence for quantum gravity that may ever be found. At this time, all that we can say is watch this space. This whole topic is very much in a state of flux at the moment, and will probably remain so for some time to come.)

Many theories of quantum gravity imply that space itself is quantized at very small dimensions, usually thought to be of the order of the Planck length. The predicted "graininess" of space is also widely believed as a consequence of quantum theory in general. This prediction is testable in theory and has recently become testable in practice also—with some very unexpected results.

The granular structure of space should leave its mark in two ways on very energetic electromagnetic radiation traversing great distances across the universe.

Assuming that space has this granular structure, as we might justifiably call it, its presence should have an effect on photons travelling through it. Imagine a level piece of ground which appears smooth at the scale of our feet, but very hilly when looked at through the eyes of an ant. We can just stride across it without taking any notice of the very small ridges and hollows over which our feet tread. However, the poor little ant does not have it so easy. It must go up hill and down dale over and over again as it crosses the same area of ground. There is no flat area at all for the ant! Similarly, if space only appears to be smooth at relatively large scales but is actually granular at the smallest dimensions, one might expect low-energy photons to pass unhindered across the hills and hollows of quantized space, whereas those photons having the highest energies would, like the ant, find the journey a far more up-and-down affair. In theory, this implies that if two photons, one having low energy and the other high, were simultaneously emitted from the same source, the one having the lower energy should in effect travel a shorter distance then the high-energy photon and arrive at the point of observation first. That is to say, the one having lower energy should skim across the hills and hollows of quantized space (just as we stride over the minute undulations of what we think of as flat ground) whereas the other must follow a path analogous to that of the ant and tackle each individual irregularity.

This should be testable observationally and, indeed, on May 10, 2009, an event was observed far away in deep space that enabled just such an observational test to be carried out. The event was the gamma-ray burst now catalogued as GRB 090510. This burst shone forth gamma ray photons covering a range of energies and by measuring precisely the times of arrival of the lower-energy rays and those of the highest energy, the maximum time delay could be determined. The very highest energy photon from the burst was shown to have arrived exactly 0.829 s after the first of the lower-energy ones.

Surprisingly, that time was too short according to all the theories. Even the most conservative calculations yielded a time gap greater than one second. To make matters worse, there is no certainty that the high-energy photon and those having lower energy were actually emitted from GRB 090510 at precisely the same instant. So the figure of 0.829 must be taken as a maximum delay only. Quite possibly, this photon was emitted later than the lower-energy ones and the real delay was shorter still. It is certainly not impossible that there was no delay at all! Nevertheless, even assuming the maximum delay, the restrictions placed on the size of the graininess of space are severe. The maximum limit of graininess permitted by these observations is 1.2 times less than the Planck length of $10^{-35}$ m. Remembering that this minimum figure is based on the conservative assumption, it remains entirely possible that there is no minimum size—no graininess—at all and that space really is perfectly continuous.

In fact, other slightly earlier observations involving the other way that graininess should leave its mark on energetic electromagnetic radiation indicate that the real limit is indeed far, far smaller. As well as having their journeys lengthened by being forced to travel up and down the tiny hills and hollows of quantized space, gamma rays should also experience a certain "twisting" (polarization) as they encounter the grains of space; those of shorter wavelength and higher energy being more affected than their lower-energy companions. In 2006, observations of the Crab Nebula using NASA's Integral gamma-ray observatory failed to detect any polarization, but that did not place very stringent limits upon the graininess of space. As a supernova remnant within our own galaxy, the Crab is relatively nearby in cosmological terms. Not so the gamma-ray burst designated as GRB041219A and previously observed by Integral on December 19, 2004. Studies of the Integral results from this object also failed to find any difference in the polarization of its higher and lower-energy gamma rays, placing an upper limit on the size of the hypothetical grains at a fantastically tiny $10^{-48}$ m! This is around ten trillion times smaller than the Planck length—and that is the maximum size that this observation allows. It is quite possible that the real size is many times smaller than this. Indeed, this result too is entirely consistent with there being no graininess at all in the fabric of space.

Not surprisingly, there are some members of the physics community who refuse to believe these results and insist that something must have gone wrong with the interpretation of the observations. Their skepticism may yet turn out to be correct, although no convincing reason why the results should not be trusted has yet been given. Assuming that the results continue to stand, the situation does not look good for the most promising models of quantum gravity. This looks like another hint that gravity may not be amenable to quantization.

It is bad enough that gravity is such a non-conformist amongst the forces of nature, in the ways already mentioned, but how and why does it alone stand out against quantization, if, indeed, this is truly so?

One possible answer—this one right out of the left field—is that it does not exist! This (at first sight shocking) proposal gets rid, in one fell swoop, of all the eccentricities of gravity plus the need for a quantum theory of this supposed force. Maybe gravity cannot be quantized because there is nothing there to quantize.

I can almost hear the reader protest "But if there is no such thing as gravity, why do apples fall from trees and planets orbit the Sun? Wouldn't we all just float away into space?" That is a fair enough comment, so here is a more complete explanation. Gravity does not exist (according to this suggestion) as a *fundamental* force of nature. It is not like the strong and weak nuclear forces and electromagnetism, which unified field theories demonstrate as having been united into a single force at high energies immediately following the Big Bang. Clearly something holds us to the Earth and keeps the Earth in its orbit around the Sun, but is this something really a fundamental force of nature? The situation may be somewhat analogous to the supposed force of suction. In the accepted carelessness of everyday speech, we might say something like "I sucked cider through a straw" without being pedantically corrected. However, if there is one thing that will make a school science teacher erupt is for one of the students to speak about a quantity of liquid being sucked up into a pipette. The writer recalls Professor Julius Sumner-Millar doubling up as if struck down by acute appendicitis on his television program when his assistant referred to something being sucked into a tube! The truth is, suction as a force does not exist. What happens when you suck cider through a straw or suck liquid into a pipette is that you reduce the air pressure in the tube below that of the atmosphere causing the latter to push the liquid up into the tube because of this difference in pressure. The liquid rises as if there is some force within the tube pulling it upward, whereas in reality it is being pushed upward by the greater atmospheric pressure outside of the tube. Could gravity be vaguely analogous to suction? May it be, not a force as such, but a consequence of something else? One controversial but very interesting and attractive theory suggests that it is.

## Gravity and the Quantum Vacuum

Earlier, we looked at the explanation of inertia given by B. Haisch and his colleagues, which tied this property to the effects of the quantum vacuum. A similar approach to gravity was suggested by well-known Russian physicist and political dissident Andrei Sakharov in 1968 and later elaborated upon by Haisch, Puthoff and Rueda who incorporated it into their wider work on inertia.

Puthoff began with the assumption that all particles are situated within a sea of electromagnetic zero-point fluctuations. As we earlier remarked, these vacuum fluctuations buffet the particle and cause it to be in a constant state of jittery motion similar to (albeit orders of magnitude less than) the Brownian motion of tiny material particles suspended in a colloid. This jittery motion is that called zitterbewegung by Schrodinger. We said all of this previously, but now comes the real reason for going over it again. In a chunk of matter, the constituent particles will not only be affected by the fluctuations within the background field, but also by the fields being generated by the surrounding particles within the same chunk of matter. This causes the particles to couple and the coupling between the particles will likewise affect the polarized vacuum in the vicinity of the chunk of matter. This will, in turn, exert a force on the jittering charged particles making up other surrounding chunks of matter. In this roundabout manner, material objects will appear to weakly attract one another. Although we name this attractive force gravity and have traditionally thought of it as a fundamental force of nature, the source of the apparent attraction is really yet another manifestation of electromagnetism. Gravity, as a separate force, is an illusion.

Haisch compares this model of gravity with the curved space model of General Relativity. Pointing out that the evidence for space being curved in the region of massive objects is inferred from the bending of light paths as they pass close to large concentrations of mass, he remarks that the same effect can also result from the rays traversing the polarized vacuum. He writes that "The warpage of space might be equivalent to a variation in the refractive index of the vacuum" analogous to the reason why a straight stick appears bent when partially submerged in a glass of water. The mathematics of General Relativity, and the strange phenomena predicted by this theory (black holes, gravitational waves, gravitational lensing and the like) remain unchanged and may equally be interpreted, either, as the curvature of space or as manifestations of the polarized vacuum. In this way, the long-awaited nuptials of relativity and quantum physics may finally take place, although the marriage might be a very different one from what the majority within the physics community may have expected. The courtship through quantum gravity would then turn out to have been a wild goose chase.

The apparent failure to find evidence of a grainy structure of space at the level of the Planck length might be seen as evidence favoring this theory to the extent, at least, that it lies awkwardly with the expectations of quantum gravity. Of course, it comes nowhere close to proving anything. Yet, the popular idea that when the universe was down to the size of a Planck length, General Relativity lapsed and quantum gravity took over and (amongst other things) neatly avoided the initial singularity predicted by General Relativity is in trouble if gravity cannot be quantized and space really does turn out to be continuous.

As a wild speculation, harking back to the earlier suggestion that physical laws come into being together with the universe itself, it might be further suggested that these laws do not pop into existence fully formed but go through a very brief evolutionary stage (probably lasting for just a fraction of a millisecond) before freezing into their present form. This may be completely wrong, but if it does contain any truth it would seem to bring both good news and bad news. The bad news is that the present laws cannot describe the very first instants of the universe, nor could any theory based upon today's physics unless the evolutionary stages of the contemporary laws can be determined. But the good news is that the present laws cannot describe the very first instants of the universe. This is good news because it would imply that General Relativity as we know it today did not operate back then and in consequence, even without quantum gravity, that pesky initial singularity may still be avoided.

## Go to Warp Speed Mr. Spock!

There are certain aspects of the line of thinking followed by Haisch and colleagues—formally known as stochastic electrodynamics (SED)—that science fiction writers love. Speculations have been voiced as to the possibility of extracting energy from the vacuum, but to date no patents have been issued for any energy from empty space inventions. Nevertheless, if the model developed by Sakarov, Puthoff and Haisch is correct, nature is essentially doing this all the time in the phenomenon that we call gravity.

But what excites certain sci-fi writers (Arthur C. Clark being one example), as well as some daringly speculative scientists, is the prospect that in principle, this theory (to quote Haisch's own words) "[opens] a door on a way of perhaps someday controlling inertia ... we had no inkling that was even possible in principle before". If there is some way (and we certainly do not know whether there is) of somehow shielding a parcel of matter—a spaceship for instance—from the effects of the fluctuating vacuum, it is arguably possible to reduce the inertia of that shielded parcel of matter and escape the limitations imposed by Relativity. By switching on such a shield, we might imagine the captain of a spaceship of the distant future propelling his craft beyond the speed of light—warp speed in the language of *Star Trek*. Or perhaps a shield will not even be necessary. If the vacuum fluctuations are confined to the brane of our universe, escaping them could be accomplished just by lifting our spaceship off the brane. "All" we would need to do is move in a direction that is simultaneously perpendicular to the three dimensions of length, breadth and height.

Coming back to reality, any thought of manipulating the vacuum remains science fiction and, moreover, even the theory on which these wild speculations are based is in itself very controversial. Yet, we should also bear in mind that travel to the

Moon was science fiction in the days of Jules Verne. Perhaps a good thought on which to conclude this section is a quote from science historian Roman Poldolny. Concerning this issue, Poldolny reminds us that "It would be just as presumptuous to deny the feasibility of useful application as it would be irresponsible to guarantee such application." That is a thought worth remembering.

## But Is the Quantum Vacuum a Real Vacuum After All!?

Just about everyone within the physics community understands the quantum vacuum as filled with virtual particles. As we saw earlier, the Lamb shift and the Casimir effect have each been derived with a high level of mathematical precision from this starting point. Moreover, such fluctuations in the background vacuum are blamed for the small degree of irremovable background interference in electromagnetic equipment and (if SED is on the right track) are even responsible for what we like to call "the force of gravity". Moreover, their presence is widely seen as a necessary consequence of the uncertainty principle of Heisenberg and as such appears to be inescapable in any quantum model of the real world.

Yet, a theory advanced around 40 years ago by Fred Hoyle and his colleague, Jayant Vishnu Narlikar concluded that this energetic vacuum of quantum theory was a myth. The vacuum really was—a vacuum; nothing in the full sense of the term!

Hoyle and Narlikar began from the work on electromagnetism by K. Schwarzschild, H. Tetrode and A. D. Fokker in the earlier part of last century. Specifically of interest was the recognition by these scientists of the phenomenon of delayed action at a distance between two charged particles. What is meant by this term is explained as follows:

Assume two charged particles (let's call them *a* and *b* for simplicity) that are both in motion along individual tracks through space and time. As they move along these tracks, they each cause disturbances in the electromagnetic fields surrounding them in accordance with James Maxwell's theory of electromagnetism.

Next, let A be a typical point on the track of particle $a$. Suppose now that a line, representing a ray of light, is extended from A in the direction of $b$. Because of the motion of $a$ and of $b$, this line will intersect $b$'s track, not at the position occupied by b at the time the light ray left A, but at a later time, and at a point which we might as well call B. This will also be true for the disturbance in the electromagnetic field generated by the motion of charged particle $a$ at point A on its track. If the two charged particles are, for example, one light hour apart, the disturbance caused by $a$ at A reaches $b$ with a delay of one hour.

That does not raise any real problems but, Narlikar points out, according to Newton's third law of motion, for every action there is an equal and opposite reaction. In this instance, the implication is that the action from A to B must be accompanied by a reaction from B to A. The odd thing about this reaction is that it must go backward in time. Therefore, if $a$ acts on $b$ via a retarded effect (that is to say by an effect moving forward in time and affecting $b$ after it leaves $a$), $b$ must also act on $a$ by an advanced effect (an effect moving backwards in time); something which is so contrary to our experience that the subject was quietly dropped soon after Schwarzschild et al. drew attention to it!

But the issue did not lie undisturbed for too long. In the mid 1940s two famous physicists whom we have already met in these pages—Richard Feynman and John Wheeler—picked it up and developed it much further. They noted that in the real universe, there are of course more than just two charges with which to contend. Disturbances from $a$ are not felt by $b$ alone, but by every other charge in the universe and likewise, these two charges receive advanced effects (or waves) from all other charges in the universe, according to this theory at least.

In an oversimplified situation, when a charge is set in motion, it generates an equal mix of retarded and advanced waves. There is complete symmetry between them, so influences going backwards in time should be on equal footing with those going forward in time, contrary to what we actually experience. However, according to the calculations of Wheeler and Feynman, when the rest of the universe and its response is taken into consideration, the effect going backward in time is neatly cancelled out, leaving only the forward response with which to contend.

The work of these physicists involved an awful lot of mathematics, but the above is the outcome of their work in a highly oversimplified nutshell.

Nevertheless, although their conclusion agrees with our experience, their model was not completely realistic in so far as they assumed a static universe, not the expanding one in which we actually live. An expanding universe complicates matters somewhat in so far as it has something which a static one does not, namely, a cosmological arrow of time. Even a steady state universe, as long as it is expanding, gives cosmological time a definite direction. However, the direction of this arrow of time can in principle be changed. An expanding universe can be changed to a contracting one by changing the sign of the time coordinate. With respect to advanced and retarded waves however, this change of sign need not be symmetrical. As Narlikar remarked "Suppose we find that the response in an expanding universe cancels all advanced waves. It does not follow automatically ... that it will cancel all retarded waves in the other solution ... simply because the transformation of time reversal ... is no longer permitted in an expanding universe." (*The Structure of the Universe*, p. 194).

Only certain models of the universe permit what Narlikar called the "correct response" of the universe, that is to say, the one that agrees with our experience. The issue of what properties the universe must possess to yield this correct response was taken up in the 1960s and 1970s by Jack Hogarth, Fred Hoyle, P. E. Roe and Narlikar himself. To make a long story short, what these scientists found was that, in order for the universe to have the "correct response," the universe of the future (the future absorber) had to be a perfect absorber (i.e. one that absorbs all radiated energy) and the past universe (the past absorber) must be an imperfect absorber. The reverse case would yield the wrong sign (cancellation of retarded or forward-going solutions) while a situation in which both absorbers were perfect would give ambiguous results.

The question as to what these properties are will be left aside for a moment. Just let's be content for the present with what we already know; namely, that the universe does in actual fact yield the correct response.

Narlikar and Hoyle then turned their attention to another mystery of nature, namely why spontaneous transitions of electrons take place (why electrons sometimes, apparently spontaneously, jump from one energy state to another, releasing a quantum of energy) and why these transitions are always to a lower energy state. Traditionally, this has been explained as due to the fluctuating field of the quantum vacuum, however Hoyle and Narlikar uncover a different possibility. They see an electron's choice as being threefold. Either it can remain at the energy state it possesses (stays in the same orbit around the atomic nucleus in terms of older atomic models), it can jump up to a higher state or it can jump down to a lower. The first choice needs no comment. The second is possible only if it acquires energy from incoming waves, i.e. from advanced waves in terms of Wheeler and Feynman. But this has zero probability of occurring in a universe having Narlikar's correct response.

The third alternative involves the possible loss of energy through retarded waves and does, therefore have a finite probability in a universe with the correct response. Thus, on this model, spontaneous transitions are explained in terms of retarded waves and are seen as further evidence that the past absorber of the universe is imperfect, the future absorber perfect and the universe to consequently possess the correct response to advanced and retarded waves. From this, Hoyle was able to calculate the exact magnitude of the Lamb shift, all in the absence of vacuum fluctuations. On this model the vacuum turned out to be literally nothing. (He did not attempt to account for the Casimir effect, as far as I am aware).

Now we return to the question of what model of the universe was found to give the correct response. It turned out to be the Steady State model. The single Big Bang gave a perfect past absorber and an imperfect future one, the cyclic Big Bang and Big Crunch yielded two perfect absorbers, as did a perfectly static universe which, however, was not a serious contender. Only the Steady State came out right.

This was one of the reasons why Hoyle never abandoned the Steady State theory, or some modified version of it, even as observational evidence for the Big Bang continued to mount. Certainly, the Wheeler-Feynman thesis, and it development by Hoyle, Narlikar et al., is attractive in the way in which it ties together the very small and the very large, the sub-atomic and the cosmic. But if it conflicts with what appears to be overwhelming evidence in favor of a universe model with which it is inconsistent, something must be wrong. The theory also has many strange ramifications and it seems that neither Wheeler nor Feynman was thoroughly convinced of its validity. At least, that appears to be the hint in Feynman's autobiographical book *Surely You're Joking Mr. Feynman*.

One of the strange conclusions of this theory concerns the necessity of an absorber before any object in the universe can emit electromagnetic radiation. Tetrode, for instance, held that the Sun would radiate neither light nor heat if it was the only object in an otherwise empty universe. Suppose that at some date in the far distant cosmic future, assuming that the universal expansion continues, the observable universe becomes empty of matter except for a single cloud that collapses into a lone star (we need not concern ourselves as to the origin of this hypothetical cloud). If Tetrode is correct, that star will not shine because there would be no other matter within its O-sphere to absorb its radiation. Yet, if there was another similar cloud a billion light years away from this star, the latter would shine because it would know, from the very moment that it collapsed into a sufficiently dense condition for its central fusion fires to start burning, that its radiation could be absorbed by another body. It would receive this information via an advanced wave that started out from the absorber cloud and travelled back in time to the radiator star.

A related bizarre consequence drawn by one writer concerns a flashlight being shone into completely empty space. It would not shine at all! If there was some matter, but not enough to absorb all of the flashlight's potential beam, the light would shine, but only feebly. The full brightness would only be achieved if there was enough matter out there to act as a perfect absorber for the flashlight's beam.

Maybe, like the differing reactions to the Schrodinger's cat thought experiment, the question of whether one believes these bizarre consequences to constitute a reductio ad absurdum of the theory or further evidence for the weirdness of the universe depends on how strange one thinks the universe can really be. And which conclusion is actually correct will depend upon how weird the universe truly is.

# The Plasma Universe

To many people, the mention of plasma brings to mind thoughts of blood. However to a physicist, the word conjures up a very different picture. Plasma, or the fourth state of matter as it is sometimes known (differentiating it from the other three states of solid, liquid and gas) is the state in which most of the visible matter in the universe exists. Basically, it could be described as an ionized fluid and manifests in ways as diverse as the orbs of stars, the finest wisps of interstellar gaseous nebulosity through to the extremely rarefied mist of ions floating between the stars. The universe is filled with plasma and, to that extent, it can truly be called a plasma universe.

That term has, however, a far more specific implication. A small group of scientists argued that plasma physics, not the more conventionally believed mechanisms of gravitation and cosmic expansion, governs the phenomena that we observe in the universe. According to proponents of this model, everything from the structure of galaxies and, indeed, of the observable universe itself, right down to the formation of stars, planets and the minor members of the Solar System (and, presumably, their counterparts in other solar systems as well) are explicable in terms of processes within plasma. Electromagnetism, they argue, is the determining force of cosmic structure, and it creates this structure through plasma phenom ena.

The principal proponent of the plasma model was Hannes Alfven, an expert in the field of plasma physics and winner of the 1970 Nobel Prize for his invention of the field of *magnetohydrodynamics* (or MHD for short). Alfven published his early ideas on the subject of plasma cosmology in his 1966 book *Worlds-Antiworlds*, and the thesis there presented was, some 5 years later, extended by O. Klein into what has come to be known as the *Alfven-Klein model* of the universe or, alternatively, *Klein-Alfven cosmology*. Whichever order one places the names, the model begins by attempting to address a continuing issue in standard cosmological theories. In experiments using particle accelerators, it quickly becomes apparent that for every new particle created, there is also a corresponding anti-particle. Every electron, for example, has its corresponding positron. Yet, the universe appears to be overwhelmingly matter. Some anti-matter is observed in certain high-energy astronomical events, but the world around us is certainly one of positive and not negative matter. This is just as well of course, as we all know what happens when positive and negative matter—matter and anti-matter—meet. It seems that, either, the initial creation process favored only one form of particle or that a small asymmetry existed between the positive and the negative and, as Einstein expressed it, matter won.

However, in the AK or KA cosmology, matter and ant-matter both exist in equal quantities in the universe, although they are kept apart by cosmic electromagnetic fields formed between two thin boundary regions consisting of two parallel layers with opposite electric charges. Interaction between these boundary regions generates radiation, which in turn ionizes gas into plasma. Both matter and anti-matter is ionized in this way, resulting in a mixed plasma to which Alfven gave the name ambiplasma. The boundary layers are therefore formed of this strange matter/anti-matter substance. Although we might think that the matter and antimatter components of the ambiplasma might explosively annihilate one another upon contact (in the typical manner of matter and anti-matter particles), Alfvin argues that the ambiplasma would actually be quite long-lasting because the particles and antiparticles composing it would be too energetic (hot) and the overall density of the ambiplasma too low to allow rapid mutual annihilation to take place. The double layers will, he speculates, act to repel clouds of particles of the opposite type while acting to combine clouds of the same type, resulting in a segregation of matter and anti-matter particles and the concentration of each into everlarger concentrations. As a further development of the model, two kinds of ambiplasma came to be recognized; heavy ambiplasma consisting of protons and anti-protons and light ambiplasma comprised of electrons and positrons.

Our section of the universe is overwhelmingly composed of positive matter because we just happen to find ourselves in one of the positive regions. Unlike conventional cosmological theories, space is not expanding. The redshift we observed really is a Doppler effect as the material within our region, not the space itself, expands outward because of the slow annihilation between the matter and anti-matter in the double layer at the boundary of our region.

There was no Big Bang in the cosmological sense, although there may have been a big bang in the sense of a great explosion in our region. These things possibly occur from time to time but, like the expansion of our region of the universe, they are just local events that are not typical of the cosmos as a whole and do not determine overall cosmic history. The wider cosmos has always existed and, as far as can be ascertained, will remain forever. The universe at large is eternal, infinite, and (on the largest scales of space and time) in a static, non-expanding steady state. The universe as envisioned by this model is about as far from that of the inflationary Big Bang as it could possibly be.

Alfvin and colleagues also used the presence of plasma processes to account for various astronomical phenomena such as the structure of galactic magnetic fields and even the shape of galaxies themselves, as well as the contraction of interstellar matter into stars and planetary systems.

The model does, however, present some very serious difficulties which have actually grown over time (always a bad sign for any hypothesis) especially as greater resolution of the CMB has been obtained. The structure, maybe even the existence, of the CMB is in strong conflict with the model and, as already noted, in good agreement with the inflationary Big Bang. CMB support for the latter grows stronger as that for the plasma universe wanes; a very significant trend.

Moreover, the plasma model cannot explain the abundance of light elements. From where does all the hydrogen come? The continuous creation of matter as proposed by Hoyle and colleagues avoided this problem for the (expanding) Steady State model that they put forward, but this avenue is not open to a static steady state universe of the type proposed by Alfven. The only recourse is to fall back on the notion of eternal matter; an idea that, we might say, went out with high-buckle shoes!

Another problem concerns the amount of gamma radiation generated by the slow but steady process of particle/anti-particle annihilation in the ambiplasma. The flux of this high-energy radiation should be far greater than what is actually observed.

The only way around this problem is to assume that our local bubble of positive matter exceeds the diameter of the observable universe. While this is possible, the suggestion is put forward on no other ground than that of simply avoiding one problem, it does nothing to answer the other difficulties with the model and is, in any case, untestable. In short, it is not a very convincing solution.

Moreover, if the universe on its largest scales is static and if it is both infinite in spatial extent and eternal in time, the old problem of Olbers' Paradox once again raises its head. Unless photons of electromagnetic radiation have a finite lifetime—for which there is no evidence—an infinite, eternal and non-expanding universe must be filled by an infinite number of photons; a conclusion that is clearly contrary to the actual state of affairs. It seems that Alfven's model leaves itself open to just such a difficulty unless something like the tiring of light is introduced purely to avoid the problem.

It is rather interesting however, to notice how both the SED model of Haisch and colleagues and the plasma model both (albeit in very different ways) dethrone gravity in favor of electromagnetism as the principal force governing the universe. This is just coincidental as there is really no commonality between these two models, but it is nevertheless interesting to note how the perception of gravity as the overriding influence of large-scale phenomena and ultimately of the universe as a whole appears to be changing— at the more radical edge of cosmological thought at least.

## The Cosmological Model of Thomas Van Flandern

Tom Van Flandern was nothing if not controversial! In his introduction to his 1993 book *Dark Matter, Missing Planets and New Comets* Van Flandern listed 38 examples of what he called "some of the most interesting points" discussed within the book. Just to give an indication of the flavor of the work, a selection of these "most interesting points" for which be argued include;

Faster-than-light motion in forward time is possible.
The physical universe has five dimensions.
The universe is infinite in all five dimensions.

There was no Big Bang.
The universe is not expanding.
Quasars are associated with our own and nearby galaxies.
A former planet exploded (just 3 million years ago) between Mars and Jupiter.
Artificial structures may exist on Mars.
Gravitational shielding is possible ... and so forth.

These are radical proposals, but Van Flandern was no crackpot. His conclusions were drawn from rational and careful arguments and, even though most of them have become increasingly difficult to accept in the face of accumulating observational evidence, even now they at least stimulate thought on the matters covered.

Van Flandern's name is probably chiefly associated with the hypothesis of an exploding planet in what is now the main asteroid belt of the Solar System, and of the asteroids as fragments of this world. Evidence in recent decades as to just how extensive and diverse the Sun's system of minor, broadly speaking, asteroidal bodies really is makes this hypothesis no longer viable, however our concern here is with Van Flandern's thoughts about the more remote realms of the universe and with more basic cosmic processes.

Like most of cosmology's dissenting voices, Van Flandern rejects the Big Bang. He also rejects the orthodox interpretation of the distance-dependant cosmic redshift of galaxies as evidence of an expanding universe. Thus far, his model of an infinite, eternal and static steady-state universe shares a certain similarity with that proposed by Alfven; a point to which Van Flandern himself draws attention. However, the similarity ends there. In Van Flandern's model, space is filled with an ether-like medium through which both gravity and light propagate via wave motion. Indeed, like the early Greek Eleatic School of philosophy (a member of which—Zeno of Elea—he quotes) Van Flandern argues that neither space nor time could exist in a void. All existence must be a continuum; everything must be connected by a medium of "substance" of which "matter" and "energy" are two manifestations. As light passes through this medium, light from distant sources loses a degree of energy and therefore decreases in its wavelength. This loss of energy, Van Flandern writes, "is as unavoidable as the loss of energy by ocean waves moving through a resisting medium of air" (*Dark Matter*, p. 93). Major disturbances will also cause waves to occur in the universal medium, and it is through these larger-scale features that matter is bundled together into the galaxy walls and similar patterns that are observed in the wider universe. He also sees Tifft's evidence for quantized patterns in the redshift of galaxies as favoring belief in this mechanism.

As already discussed, serious problems arise in any "tired light" model of this sort and the evidence which at the time Van Flandern was writing seemed at least plausible has since been weakened by further data (the alleged quantized redshifts being a good example of this). But the most radical proposal by Van Flandern is not directly related to his theories of redshift or even of cosmic non-evolution. It concerns the dimensionality of space.

Van Flandern understands the universe as possessing five dimensions, each of which is infinite in extent. Four of these are universally agreed upon—length, breadth, height and time—although certainly not everyone would hold that they are infinite in extent. The fifth dimension, according to this model, is scale. Like the other dimensions, this too is infinite. Looking downwards (so to speak) the range of scales disappears into what, for us, is the infinitely small, while looking in the other direction, scales keep increasing to larger and larger volumes. Van Flandern introduces the axiom that "The universe should look essentially the same at all scales" but without implying that the set of finer details experienced at our scale is necessarily representative of the universe at large—either at our scale or at any other. He writes;

> The axiom is not meant to imply that blocks of space, air, land and water would look essentially the same. Rather, it implies that such forms are not unique to our scale, and similar structures will be found at vastly different scales. Conversely, the large structures we see at our scale, such as stars, galaxies, walls [of galaxies] and voids are not necessarily the same structures we might find in different parts of the universe at our scale. In this connection, it may be sobering to appreciate that there are as many atoms in a single drop of water as there are stars in all the galaxies in the visible universe. Yet just outside the single drop of water, the structure of the universe changes drastically – for example, to fast-moving air molecules. (*Dark Matter*, p. 26).

Nevertheless, the forces that are dominant on each scale may vary. For example, on the scale of the atomic nucleus, the strong nuclear force is dominant while gravity is, for all practical purposes non-existent. On our scale, gravity is important. On some vastly larger scale, some force which has no more effect at our scale than gravity does at the scale of a proton may be the dominant one and gravity itself may play no part.

This model brings to mind the rhyme;

Little fleas have lesser fleas upon their backs to bite 'em
And lesser fleas still lesser fleas, and so ad infinitum!

Van Flandern's mention of atoms in a drop of water also reminds us of a suggestion occasionally heard that perhaps our whole universe exists on a microscope slide in some super-universe, and that right now we are being examined by a physics professor with eyelashes billions of light years long! Let's hope he does not decide to wash us down the drain ... although on our scale of things we would probably not even notice if he did! And then, of course, he might also be under examination by another professor with even longer eyelashes ... and so on ad infinitum.

I am not saying that Van Flandern would have drawn such fantastic conclusions as these. Even such an adventurous soul as he must draw the line somewhere! Indeed, even the weird universe must draw the line of weirdness somewhere. We might like to let our minds drift over such fantastic notions while engaged in some monotonous task, but serious speculations they are certainly not!

More seriously, the notion of a dimension of scale in Van Flandern's sense seems an unnecessary complication and one which is not testable in any scientific sense. Like the plasma model, Van Flandern's is also just too far outside of the field. That is not a criticism in itself, as most of what we now accept was once outside the field of the acceptable. However, the difference between models that were once scientific heresy but are now orthodoxy and those that are likely to remain heresy is in the last analysis determined by evidence. At present, the evidence increasingly supports the inflationary Big Bang and continues to weaken static and/or steady-state alternatives such as the models of Alfven and Van Flandern. It is possible that this will change someday, although the chances of this happening are also decreasing as data increases. Although the present version of the inflationary Big Bang will probably change in the future (just as it has superseded the Gamow model which in its turn superseded Lemaitre's original version) most cosmologists would be surprised to see a truly revolutionary change in our understanding of the universe. But three cheers for those non-conformists whose attempts at just this help to keep the rest of the cosmology community on its collective toes.

## The Complex Theory of Relativity: A Universe of Eons

This original theory was proposed by physicist Jean Charon, whom we already met in our earlier discussion of time dilation, space contraction and the so-called twin paradox. As we saw there, his thinking on these subjects was somewhat outside the square of strict orthodoxy, but his penchant for original thought blossomed even further with the 1977 publication of his Complex Relativity Theory. The "complex" part comes, not from the complexity of the theory itself, but from the necessary use of complex numbers in its highly mathematical formulation. A complex number is one that has both real and imaginary components and can be written in the general form $x+iy$, where both $x$ and $y$ are real numbers and $i$ is known as the imaginary unit. The imaginary unit is equal to the square root of $-1$ (i.e. the square of $i$ is equal to $-1$). The real part of the complex number is $x$ and the imaginary part is $iy$. The latter is defined as an imaginary number because it is a multiple of the square root of $-1$ (Fig. 5.7).

The mathematics need not concern us, but the broad gist of Charon's theory is his postulation of the existence of very small time-spaces, effectively little closed universes present throughout the larger universe. These are abundant both in all matter and throughout space. They were born in the first few instants of the Big Bang (and can be viewed as, in a sense, sparks of the Big Bang itself) and are present in the universe as stable structures all around, and indeed within, us. Charon names them eons.

FIGURE 5.7 Jean Charon 1920–1998 (*Credit*: Gerrit Teule)

Mathematically, both these eons and the space in which we live can equally be viewed as spherical. Now, spherical spaces, irrespective of their size, can only touch each other at point-like locations. This fact forms the crucial basis of Complex Relativity, namely, the time-space of every eon touches our space at just one point-like location. Therefore, to the degree that they are observable at all in our space, eons manifest as point-like entities. These point-like particles, Charon identifies with electrons. From one point of view therefore, an eon may be seen as the extension of an electron beyond our three-dimensional space while from an opposite but equally valid viewpoint, an electron may be thought of as the shadow or fingerprint of an eon imprinted upon the spatial realm of our familiar universe.

Each eon is filled with a cloud of black body radiation, revolving photons and a rotating neutrino. Eons likewise possess extremely high and pulsating temperatures and densities. Indeed, so high is the photon concentration within an eon that space is warped around it in a way that is reminiscent of a black hole. Eons, however, are not black holes according to Charon, even though they do warp space in a similar manner. This is how each eon becomes a miniature universe in its own right—a little spark of the Big Bang in which the primordial temperature and density of the Big Bang is preserved. In fact, it is preserved unchanged forever because each eon is a completely closed mini-universe in which this energy is eternally trapped. The law of entropy does not apply within an eon; the informational order therein can only remain stable or even increase as photons within the eon interact with one another to form new informational configurations. What it can never do is decrease.

When understood as the points at which eons touch space, some of the strange properties of electrons can be viewed in a different light. Thus, if the mass and the charge of an electron originate in the eonic extension of the particle, the mystery as to how a point-like entity can have both mass and charge is no longer so mysterious. The point appearing in our space and time is no longer all that there is to an electron. The greater portion of each eon lies not in our familiar three-dimensional space but in what Charon calls the imaginary dimension. This is not imaginary in the sense that it is unreal or exists only in the mind, but in the sense that it is explicable only in terms of the mathematics of complex numbers and their imaginary number components.

Because they were all created together in the Big Bang, eons and the photons within them are quantum entangled and, Charon argues, can act non-locally upon photons in other eons irrespective of their distance. In their spatial manifestation as electrons, eons can start chemical reactions and, through non-locality, communicate with other electrons in a way that affects the behavior of the latter.

On the basis of Complex Relativity, Charon became the first person to actually calculate the charge of an electron rather than simply accepting it as a given fact of nature. His answer agrees with observation to an accuracy of 1.9 %, which he regarded as strong support for the validity of his theory.

In later work, Charon becomes more speculative and extends the scope of eonic theory into the mental and spiritual realms. Strongly influenced by Teilhard de Chardin, Charon saw the information-storing qualities of eons as a means of accumulating information throughout the evolutionary process. Furthermore, because each electron in every atom of our body and brain is the point at which an eon touches our space, and because they are all quantum entangled, non-local communication between the eons within our body and even between our body and the bodies of others around us is presumably taking place all the time. This communication between eons is not only the basis for mind (when it acts within one's body and even between bodies) but, in the wider universe, manifests as a cosmic tendency toward that which Chardin termed complexification. Life and ultimately conscious intelligent life, is therefore in the last analysis determined by the properties of eons.

Not entirely dissimilar thoughts concerning the emergence of fully-developed mind from some more rudimentary proto-mental manifestation have been voiced by other scientists over the years. Julian Huxley, in his *Essays of a Humanist*, speculated that mentoid properties might belong to individual cells and that, when these come together in a living organism of sufficient complexity, mind emerges. Charon locates mentoid qualities (although he does not use this term) in the even more basic structure of the eon. He hoped that ultimately the eon thesis would not only be seen as the unifying foundation for the physical forces of nature, but that it would also provide a sound basis for a satisfactory theory of mind itself. His physics colleagues were not so enthused apparently and the model has been little discussed except by those already drawn toward something approaching Teilhardian philosophy. Though an ingenious and radical proposal, the eon hypothesis remains on the left field at this time, while its supporters continue to await its day in the Sun.

## Monster Particles of Little Mass?

Neutrinos are funny little critters. They have practically no mass, no electric charge and very little size. All day and every day, our Earth is being constantly bombarded by neutrinos from deep within the Sun and in 1987 an intense shower of these particles passed through our planet from the supernova in the Large Magellanic Cloud. Neutrinos react so weakly with other particles that a beam of them could pass straight through the Earth will little diminution and, for them, small objects such as human bodies scarcely exist. Billions of solar neutrinos pass through each of us every second, yet we experience nothing of this and the neutrinos themselves don't register that we are even here.

But if we think that modern neutrinos are weird, spare a thought for those created during the first instants of the Big Bang, at least according to George Fuller and Chad Kishimoto of the University of California in San Diego. According to their study, such relic neutrinos may have been stretched by the expansion of space until they now have diameters of about 10 billion light years, close to 20–25% as large as the actual current radius of the observable universe.

According to quantum physics, the size of a particle is defined as the range of possible locations of that particle and it is this "range" that has expanded enormously, according to these physicists.

This prospect has raised the specter of objects describable in terms of quantum theory now taking on dimensions larger than anything previously known except, that is, for the universe in its entirety. How do these meganeutrinos react with, say, the gravitational pull of a galaxy or cluster of galaxies? Nobody knows.

Despite their expected size, these megaparticles will not be easy to find. Although the universe should be full of them, each is now expected to have an energy of just one ten-billionth, or thereabouts, that of a solar neutrino; and we know how difficult it was to satisfactorily detect the full quota of these! Yet, according to Fuller, because of their number, these particles should collectively have enough mass to exert a significant gravitational pull and, through this, to influence the evolution of the universe's large-scale structure. They may be a contributing factor to the dark matter of the universe. Maybe the detection of meganeutrinos will yet come about through the detailed analysis of suitably grand structures in the universe.

## Yet More Gravitational Weirdness?

There is something decidedly odd about the way stars orbit the centers of galaxies. We all know about Kepler's Second Law which, in effect, says that the closer an object is to the center of mass around which it is orbiting, the faster it will move. In the Solar

System, to take an example, Mercury moves faster than Earth and both of these inner planets move faster than distant Saturn. Yet, when the velocities of stars orbiting the centers of galaxies are measured, a very curious fact emerges. They appear to violate Kepler's Second Law. At least, after a certain turnover point is reached, they seem to leave this law behind. Travelling outward from a galactic center, the innermost stars have the highest orbital velocity (as expected) but when a certain distance from the galactic core is reached, orbital velocities cease to decrease with distance, but instead form a sort of plateau that persists right to the outer fringes of the galaxy. This odd phenomenon is known to astronomers as the "flattening of galactic rotation curves" and is generally explained by the presence of a halo of "dark matter" surrounding the visible material of galaxies. Dark matter, as already mentioned, seems to constitute a sizable percentage of the universe and has been named as the culprit in several cosmological mysteries. Still, in spite of several hypotheses as to what constitutes this mysterious stuff, a direct detection of dark matter is yet to be made.

An alternative hypothesis was put forward in 1983 by Mordehai Milgrom of the Weizmann Institute in Israel. What Milgrom proposed amounted to a modification of the theory of gravity. Not a modification of the theory of what gravity is (unlike Haisch, for instance), but a modification of the way in which it acts. Whether the nature of the gravitational attraction between objects is interpreted along the lines of Newton, Einstein or Haisch, it has always been universally agreed that Newton was right in understanding this to be directly proportional to the product of the masses of the objects and inversely proportional to the square of the distance between them. What Milgrom suggested is that gravitational attraction is not simply dependent upon the mass of a body but upon a more complex relationship involving both the mass and the acceleration of that body due to gravitational force. In practical terms, this would mean that as the orbital velocity of stars in a galaxy fell with increasing distance from the center, a point would be reached where the predictions of unmodified Newtonian gravity, and Kepler's Second Law which is governed by it, no longer holds. In short, he was able to show that his theory of

*Modified Newtonian Dynamics* (MoND) is capable of explaining the flattening of galactic rotation curves without invoking the presence of dark matter.

Not surprisingly, reactions to MoND have been mixed, in no small part to the confusing fact that predictions made on its basis have likewise been mixed!

For example, before any rotational curves of a class of objects known as low surface brightness galaxies (LSBs) had been determined, Milgrom predicted that, because of the diffuse nature of these objects—which more closely resemble the outer fringes of typical galaxies than their compact inner regions—almost all of their stars should lie within the flat or plateau section of their rotation curves. Subsequent measurements of the rotation curves of LSBs has shown that these agree very well with Milgrom's predictions.

In general, predictions of the rotation curves of galaxies of all classes—not just LSBs—based on MoND turn out to be very accurate. This may be seen as a very positive note. However, predictions of the structure of cosmological features larger than galaxies, such as galaxy clusters and the even larger structures found in recent decades, are not as good and tend to favor dark matter. For instance, MoND-based predictions concerning the motion of galaxies around the centers of galaxy clusters yield results that are inconsistent with observation.

A problem with MoND for scientists who like their hypotheses to have a strong theoretical basis, is the apparent lack of any real fundamental reason for believing it. It is essentially a proposed theory set to explain unexpected observational data. It has not been derived from some broader theoretical perspective. Lee Smolin et al. attempted to find a theoretical basis for it in the (itself uncertain) theory of quantum gravity. This project appears to have come to nothing.

MoND is certainly an interesting idea, but one which at this present time has been neither proven nor disproven. If a good candidate for dark matter is found, MoND will no doubt be resigned to that place where all good theories go when they die. On the other hand, if dark matter remains elusive, MoND may yet see its day in the Sun.

# The "Magic Number"; Deep Mystery or Just a Coincidence?

One weird feature of the universe cannot go unmentioned. It may be nothing more than an odd coincidence, but it has puzzled a number of physicists for quite a few years and has led to several unconventional speculations as to the nature of the universe and the relationship between several basic cosmic parameters. For these reasons alone, it deserves at least a mention.

The issue in question is the frequent appearance of certain numbers; certain very, very, large numbers.

One that has drawn special attention is $10^{40}$ or, in more popular notation, the figure one followed by forty zeroes! This number, some close approximation of it, or its square, cube or square root keeps cropping up in all sorts of places in physics and cosmology. It is sometimes dubbed the magic number. Let's look at some examples:

Gravity is weaker than the strong nuclear force by a factor of $10^{39}$ near enough to the magic number when such magnitudes are being spoken about.

The universe contains about $10^{80}$ atoms, i.e. the square of $10^{40}$.

The volume of the universe expanded by a factor of $10^{78}$ (approximating the square of $10^{40}$) during the inflationary era.

The diameter of a proton is $10^{20}$ (the square root of $10^{40}$) times greater than the Planck Length.

The relation $ct/r_e$ (where $c$ is the speed of light, $t$ is the age of the universe and $re$ is the classical electron radius) approximates $10^{40}$ in units where $c=1$ and $re=1$.

The ratio of the electrical force acting between a proton and an electron to the gravitational force between them is given as $0.23 \times 10^{40}$.

The ratio of the range of the strong and gravitational forces is $4 \times 10^{40}$.

The difference between the observed value of quantum vacuum energy and that calculated by Quantum Field theory comes out at $10^{120}$ i.e. the cube of $10^{40}$.

The ratio of the mass/energy in the observable universe to the energy of a photon having a wavelength the size of the observable universe is $8 \times 10^{120}$.

A typical electromagnetic particle is, according to B. Sidharth, a collection of $10^{40}$ Planck oscillators and the universe is a collection of the cube of this number. (A Planck, or radiation, oscillator is an oscillator which can absorb or emit energy only in amounts that are integral multiples of Planck's constant multiplied by the frequency of the oscillator).

There are also some other more contrived relationships involving the "magic number". For example, the age of the universe in terms of the Planck time (i.e. the time it takes for a photon of light to travel one Planck length in vacuo) approximates the cube of the square root of $10^{40}$ and the volume of the observable universe in terms of Planck volumes (i.e. the cube of one Planck length) is close to the cube of the age of the universe measured in Planck times, i.e. is near the square root of the "magic number" raised to the ninth power. (The actual value is given as $4 \times 10^{185}$, but five orders of magnitude, whilst large in our familiar terms, is tiny when compared with the magnitudes considered here).

Actually, these last look a trifle less contrived if the chief focus of the "magic" is shifted from $10^{40}$ to another very large number, viz. $10^{20}$. As the square of the latter, the former retains a degree of "magic", but taking the latter as the real base draws attention to even more odd coincidences. Moreover, it could be argued that because it measures the diameter of a proton in terms of Planck lengths, there is something quite fundamental about $10^{20}$ and that this may give it an even greater degree of interest than its famous square. After all, the proton is the nucleus of an atom of hydrogen which in a sense could be called the "primal element" of the universe.

For no other reason than convenience, let's denote $10^{20}$ as $N$. We may then write:

$N$ = Diameter of a hydrogen nucleus in Planck lengths.

$N^2$ = Difference between strong force and gravity etc. (See above list for further relationships).

$N^3$ = Approximate age of the universe in Planck times. It is also the difference between the mass of a typical star and that of an electron (after P. Jordan), the approximate diameter of the universe in Planck lengths, the approximate reciprocal of Hubble's constant in Planck units and the approximate mass of universe in terms of Planck mass (a value equal to $2.12 \times 10^{-8}$ kg). The square root of

$N^3$ (i.e. $N^{1.5}$) is also close to the reciprocal of the temperature of the present-day universe in terms of Planck temperature (the Planck temperature being $1.41 \times 10^{32}$ K or the temperature of the universe when its age was just one unit of Planck time).

$N^4$ = Approximate number of atoms in the universe.

$N^6$ = Difference between observed and predicted values of vacuum energy. Approximate ratio of mass/energy in observable universe to the energy of a photon having a wavelength equal to that of the observable universe. Approximate value of the reciprocal of the Cosmological Constant.

$N^9$ = Approximate volume of the observable universe in terms of Planck volume.

Furthermore, measurements by V. Shemi-Zadah of various cosmological entities in terms of the Planck length has also turned up several numbers of the order of $10^{60}$, i.e. $N^3$. And it may even be worth reminding ourselves that the usual estimate of the age of the universe at the end of the inflationary epoch approximated the square root of $N$ as measured in units of Planck time. This last may, however, end up as a spurious correlation if the primordial B-mode polarization results announced in March 2014 are correct. The unexpected strength of the signal might indicate an earlier time for the inflationary epoch according to preliminary analysis; maybe right back to the end of the Planck era itself.

The big question is whether this is hinting at something important or whether the whole thing is all co-incidence. At one level, it exudes the whiff of numerology or something similar. Is there any difference between relating the diameter of a proton to the age of the universe and finding some correspondence between the height of the Great Pyramid and the age of the US President or something equally absurd?

Some see little difference, however other scientists are not so sure. One difference, which may be important, is the use of natural units in measuring these values. It is one thing to find apparent relationships between sets of arbitrary scales (inches and minutes and the like), but what can be said when we find apparent relationships between Planck values? Is something real, and deeper, implied here? It is interesting to note, in this context, that one second approximates $10^{43}$ units of Planck time. This value would be close enough to $10^{40}$ to be included as another large-number

FIGURE 5.8 Paul Dirac 1902–1984 (*Credit*: Nobel Foundation)

coincidence if seconds were also fundamental units. Being arbitrary however, this relationship is surely just a coincidence and trying to read too much into it really would drift into the realm of numerology. Or does this example support those who say that this is all numerology anyway?

One physicist who thought that the large-number coincidences really do speak of some deep property of the universe was P. Dirac. Dirac thought that these co-incidences (at least, those known in the 1930s when he worked on this issue) hinted that the so-called constants of the universe were not strictly speaking constant at all, but that they actually evolved over time. Working from the apparent large-number relationships, he proposed that the strength of gravity is proportional to the age of the universe and that the mass of the universe is proportional to the square of the universe's age. The values of these are not, therefore, independent and their apparent relation is therefore not coincidental (Fig. 5.8).

Others have followed Dirac in this respect and proposed a universe of evolving "constants", usually within a Machian-type framework. Others, exemplified by Robert Dicke, have argued that these numbers are a necessary co-incidence for the existence of intelligent life, as different values of the parameters quantified by them would lead to different astrophysical processes than those that now take place in the cores of stars. We know that the elements essential for the existence of life are synthesized through the processes taking place in stars, but in a universe where these processes were replaced by different ones, complex organisms probably could not exist. But just why the parameters that make for such a life-friendly universe have the observed values and why these values appear to be related in the manner in which we find them to be, is not at all clear.

Positions similar to those taken by Dirac and Dicke meet with much skepticism amongst scientists, although some of the more adventurous souls amongst the physics community have toyed with similar ideas from time to time. Fred Hoyle, for instance, opined that the laws of physics and the alleged constants of nature may be different in other parts of the universe and a few other scientists have taken an evolutionary view similar to that of Dirac. The problem is, there is no evidence to suggest that the laws and/or constants of nature were any different in the earliest galaxies than they are in the contemporary universe. For such a radical view to be accepted, observational evidence must be very strong. All available evidence, however, points in the opposite direction. As far back as we can ascertain, the laws of physics and the constants of nature appear to be fixed. Maybe, as suggested earlier, there was a very brief period when the age of the universe was close to one unit Planck time, that the laws of physics were fluid, but even if that very wild speculation is true, it is unlikely to explain the strange large-number coincidences. Perhaps we are simply trying to read too much into what may after all be nothing more than a mere coincidence. Or perhaps there is, in the last analysis, no such thing as a mere coincidence. In this weird universe can we be sure which is correct?

# Appendix: The Tripple Alpha Process; An Example of Remarkable Fine-Tuning

Stars like the Sun acquire their energy from the fusion of hydrogen into helium; the same process that threatens the destruction of civilization through the hydrogen bomb or maybe helps to preserve it via controlled thermonuclear fusion. As it occurs in the Sun, it is the process that enables life to flourish on Earth.

But the universe consists of much more than hydrogen and its fusion product, helium. Other elements must also be built, albeit in places even more extreme than the core of the Sun. Some of these other places are the central regions of stars that have aged well beyond the Sun's years and have evolved (or, we may equally say, "decayed") to a point beyond the hydrogen burning (fusing) stage of their lives; beyond what astronomers call the Main Sequence. These relatively massive stars are literally running out of available hydrogen and, as the thermonuclear reactions that supply the energy to maintain the star's internal pressure wane, their cores start to contract. As they leave their Main Sequence existence, their cores contract and grow even hotter, causing their outer layers to bloat into enormous envelopes having diameters equivalent to those of the orbits of the inner planets of the Solar System. For example, a star like the Sun is expected to bloat until its surface will almost reach the orbit of Earth. More massive stars at this advanced stage of their lives can become so large that even a planet as remote as Jupiter would be swallowed up by their distended globes. Because the outer regions of these stars are relatively cool, they appear in our skies as yellowish-red objects and for this region are known as red giants.

Just because their outer layers are cool by stellar standards does not mean that their cores are also cool. Far from it. Deep within these stars temperatures soar beyond anything found even at the center of our present-day Sun. When these temperatures exceed about 100 million Kelvin (at least 5.5 times hotter than the core of the present-day Sun) the helium that has built up there as "ash" from many millions of years of hydrogen burning is given sufficient energy to begin its own fusing. Unlike the outer reaches of these distended stars, pressures are high within their cores and helium atoms—or rather their nuclei stripped of electron shells—are packed in close together.

A helium nucleus consists of two protons and two neutrons. These denuded atoms of helium are known as alpha particles. Now, the fusion of two alpha particles should result in an atomic nucleus consisting of four protons and four neutrons, in other words, the nucleus of an atom of Beryllium-8 (Be-8). Amazingly, the ground state of Be-8 has almost exactly the energy as the excited state of two alpha particles. The energy (temperature at the red giant's core) required to fuse two alpha particles into a single Be-8 nucleus essentially equals the energy required to break down a Be-8 nucleus into its component alpha particles. It is almost as if Be-8 should not exist at all. In fact, it almost doesn't. The half-life of Be-8 is just 0.00000000000000007 seconds. Compared with this, a lightning flash seems like an eternity, but it is nevertheless long enough for something remarkable to happen, as we will see in a moment. It so happens that the excited state of Be-8 is also essentially a match in energy to the excited state of Carbon-12 (C-12). These resonances greatly increase the otherwise negligible probability that a third incoming alpha particle will combine with the Be-8 nucleus during its incredibly short lifetime. This possibility—still very remote—is also helped by the density of alpha particles in the star's core. Many alpha particles will be fusing into Be-8 at any instant, so there will always be a good supply of the latter in spite of the very brief duration of the existence of each nucleus. The situation has been compared with running a stream of water through a sieve. Normally, the sieve will not hold water, but if the flow is sufficient, at any instant there will be a small equilibrium amount of water in the sieve. Similarly, there will at any instant be a small equilibrium concentration of Be-8 nuclei in the star's core and some of these will pick up an extra alpha particle to become C-12. In its excited state this is also unstable and most of the carbon nuclei will quickly decay, however an occasional one will happen to emit a gamma-ray before it has a chance to decay and, by doing so, attain its stable ground state.

As a side process, there is a small probability that a carbon nucleus formed by the main process will combine with a fourth alpha particle to produce a stable oxygen isotope. The process, however, largely stops there. The next step would be the addition of yet another alpha particle to form a neon nucleus from that of carbon; however this is extremely unlikely to happen thanks to the rotational properties of the carbon nucleus. The result is that carbon and oxygen, rather than neon, is produced in relative abundance in the cores of red giant stars, subsequently being distributed into space either through the slow dispersal of more modest red giants in the form of diffusing planetary nebulae or more violently as the larger members of the tribe explode as core-collapse supernovae. From there, these elements become incorporated into planetary systems and, eventually, living organisms on at least one planet of which we are aware. It is a sobering fact that without this incredibly unlikely process, there would be no life anywhere in the universe—ever.

This triple-alpha process, as it is now known, was discovered by Fred Hoyle and as he contemplated the very tiny probabilities of such a thing even happening in nature, he was driven to remark that it is "a put up job!" Atheist though he claimed to be, Hoyle nevertheless concluded that some super-intelligence must have had a hand in arranging the fundamental laws of physics allowing such a process to occur and thereby making possible the existence of living beings.

A counter claim was made that if an intelligence was capable of such a feat, would it not have been better to have arranged for Be-8 to have been more stable and to have allowed for a more productive rout to the fusing of carbon and oxygen? Maybe just giving Be-8 a longer half-life would have been sufficient.

If that had happened, these elements would be far more common and, so this argument runs, life far more abundant than it appears to be on a cosmic scale.

It seems to me that this last conclusion does not follow.

A shorter half-life (slower decay rate) of Be-8 would lead to increased fusion of heavy elements and catastrophic explosion of stars reaching this stage of their evolution. Increased numbers of exploding stars and rapid enrichment of the interstellar medium with heavy elements does not really make for a more life-friendly universe.

But suppose carbon was capable of some more direct and easy synthesis and suppose that the universe was far richer in this element than it is today. Planets like Earth would be a lot more carbon rich, but I do not think that this would increase the possibilities of life-friendly worlds. Quite the opposite in fact. What would be more frequently encountered would carbon-rich worlds with cores of diamond. As discussed in *Weird Worlds*, diamond planets are far from being a girl's best friend … or any living organism's best friend for that matter. Planets having diamond interiors lose their internal heat rapidly which, together with their rigid cores, means that plate tectonic activity of the type responsible for recycling the surface of Earth, as well as ensuring the maintenance of protective magnetic fields shielding their surfaces against energetic particles from their suns, is absent. The carbon-rich planets discovered in the real universe are not considered likely homes for life. In a more carbon-rich universe than our own, these planets would be more representative of small worlds in general than they are in the actual universe. Far from being more prevalent in such a universe, life might be absent altogether.

In short, the ease or otherwise with which carbon and oxygen are produced may itself be quite finely tuned. If the triple-alpha process was not possible, our universe would be lifeless, but if it was replaced by a process that increased the production of these elements, life might equally be impossible. Makes for interesting speculations, do you not agree?

# Author Index

**A**
Alfven, H., 242–246, 248
Alpher, R., 15–17, 45
Anaximander of Miletus, 1
Anaximenes of Miletus, 1
Anderson, A., 1–3, 5, 132
Anderson, J., 1
Aristotle, 3, 6
Arp, H., 38, 215, 216
Augustine, St., 133

**B**
Barr, S., 137
Barrett, J., 191
Barrow, J., 143, 150
Bekenstein, J., 107, 108
Bell, J., 195, 200–202
Blake, W., 6, 20
Bohm, D., 173–175, 191
Bohr, N., 163–168, 171, 172, 174–177, 180–190, 192
Bolton, J., 35
Boltzmann, L., 106–108
Bondi, H., 27, 29, 42
Born, M., 168–175
Brown, R., 161
Burbidge, G., 31, 224
Burbidge, M., 31, 224
Bush, G., 118

**C**
Calder, N., 123
Carol, L., 24
Casimir, H., 63, 99, 136, 237, 240
Charon, J., 75–78, 249–252
Chown, M., 174, 181, 188
Christodoulou, D., 107
Clark, A., 236
Coleman, J., 154
Copernicus, N., 7, 42
Croasdale, M., 228

**D**
Darwin, C., 158
Davies, P., 63
De Broglie, L.-V., 64, 172, 173, 175, 191, 192
de Chardin, P.T., 150, 251
Democritus, 158
Dicke, R., 259, 260
Dirac, P., 170–172, 259, 260

**E**
Eddington, A., 80
Einstein, A., 10, 27, 30, 56, 60, 64, 68–72, 77–80, 83–85, 104, 158, 161, 163, 166, 167, 172, 175, 184, 194, 195, 200, 201, 243, 254
Elliot, H., 158
Everett, H., 189–193
Ewing, A., 88

**F**
Feynman, R., 57–68, 238, 240, 241
Firsoff, V., 20–22, 24
Flew, A., 72, 151–155
Fokker, A., 237
Fowler, W., 31
Fuller, G., 253

**G**
Galileo, G., 7, 79
Gamow, G., 15–18, 27, 29, 30, 32, 33, 45, 249
Gardner, M., 151
Gold, T., 27, 29, 42, 81
Guth, A., 49–51
Guthrie, B., 228

**H**
Hafele, J., 72, 78, 81

# Author Index

Haisch, B., 62–68, 140, 234, 236, 245, 254
Halley, E., 125
Hardy, L., 191, 192
Hawking, S., 68, 98–101, 107, 134, 136, 202
Hazard, C., 34, 35
Heisenberg, W., 51, 63, 98, 108, 137, 140, 168–176, 181, 194, 195, 237
Heraclitus, 3
Herman, R., 17, 27, 45
Higgs, P., 66–68
Hilbert, D., 131, 132
Hoffding, H., 182
Hoffmeister, C., 42
Hogan, C., 109, 110
Hogarth, J., 239
Holba, A., 228
Holman, R., 203
Howard, D., 181
Hoyle, F., 27, 29, 31, 33, 38, 40–42, 68, 140, 149, 150, 152, 237, 239, 240, 244, 260
Hubble, E., 12, 20, 104, 217, 257
Huxley, J., 252

## I
Ijjas, A., 53

## J
Jeans, J., 24, 25
Jordan, P., 169, 170, 257

## K
Kant, I., 183–185, 188
Keating, R., 72, 78, 81
Kishimoto, C., 253
Klein, O., 242

## L
Lamb, W., 63, 99, 136, 237, 240
Laplace, P., 88, 89, 93
Lawrence, D., 6
Leibniz, G., 183, 184
Lemaitre, G., 13–15, 18–19, 22, 23, 26, 29, 249
Leucippus, 158
Levin, J., 116, 131
Lovell, B., 32
Lyttleton, R., 32

## M
Mach, E., 60–62, 65, 69, 158, 161
Maldacena, J., 202
Mandl, R., 84, 87, 220
Marsden, E., 164
Matthews, T., 34
Maxwell, J., 157, 237
Merrill, P., 31
Mersini-Houghton, L., 123, 202, 203
Michell, J., 87
Milgrom, M., 254, 255
Milne, E., 22–24, 27

## N
Napier, W., 228
Narlikar, J., 40, 237–240
Newton, I., 59, 78, 157, 163, 184, 254
Nietzsche, F., 130

## O
Olbers, H.,

## P
Paal, G., 228
Paczynski, B., 83
Pagels, H., 146, 158, 185, 190, 193, 195–198, 200, 201
Pauli, W., 171
Payne-Gaposchkin, C., 15
Penrose, R., 53
Penzias, A., 44, 45
Perrin, J., 161
Picard, C., 116–118
Planck, M., 52, 53, 55, 56, 104, 108–110, 115, 118–120, 122, 132, 135, 159–161, 163, 166, 170, 175, 203, 230, 232, 235, 256–258, 260
Plato, 127
Plutarch, 4
Podolsky, B., 194
Poldolny, R., 237
Pope, A., 5, 151
Ptolemy, 6
Pusey, M., 191
Puthoff, H., 62–64, 230, 234, 236

## R
Register, B., 182
Reuda, A., 62

Roe, P., 70, 239
Roemer, O., 70
Rosen, N., 194, 201
Rudolph, T., 105, 191
Russell, B., 20
Rutherford, E., 163–168
Ryle, G., 28, 155
Ryle, M., 32

**S**
Sakharov, A., 234
Sandage, A., 34
Sargent, W., 225, 226
Sazhin, M., 103
Schild, R., 105
Schmitt, J., 42
Schrodinger, E., 64, 65, 172–179, 181, 188, 189, 192, 194, 234, 241
Schwarschild, K., 93, 94
Sciama, D., 32, 61
Seyfert, C., 222–225
Shannon, C., 106, 107
Shemi-Zadah, V., 258
Sidharth, B., 2557
Silk, J., 116
Slipher, V., 8–10, 12
Smolin, L., 255
Socrates, 6
Sorkin, R., 108
Steiner, F., 116
Steinhardt, P., 208
Stephan, E., 219–224
Sumner-Millar, J., 233
Susskind, L., 102, 108, 201, 202

**T**
't Hooft, G., 108
Taylor, R., 153–155
Tegmark, M., 121, 125, 126, 190
Tetrode, H., 237, 241
Thales of Miletus, 1
Tifft, W., 227, 228, . 247
Tipler, F., 143, 150, 151, 155
Turok, N., 208

**U**
Unrich, W., 63

**V**
Valentini, A., 191–193
Van Flandern, T., 245–249
Veldman, M., 67
Verne, J., 237
Vorontsov-Velyaminov, B., 224

**W**
Walker, D., 162
Wheeler, J., 88, 106, 107, 190, 191, 238, 240, 241
Wigner, E., 150, 178–180, 188, 189
Wilson, R., 44, 45
Witten, E., 68, 205

**Z**
Zeno of Elea, 246
Zwicky, F., 19

# Subject Index

**A**
Absorber
  future, 239, 240
  imperfect, 239
  past, 239, 240
  perfect, 239–241
Adiabatic process, 5
Agnosticism, 184
Air (as primary substance), 3–5, 12, 41, 68, 80, 147, 233, 247
Alpha particles, 164
Ambiplasma, 243, 244
Anthropic principle, 143, 150, 151
Apollo Moon landings, 28, 86
Atoms, 9, 14–17, 24, 25, 30, 45, 91, 99, 126, 130, 157–211, 247, 248, 251, 256–258
Axis of evil, 118, 120–123, 203

**B**
Bell's inequality, 197–200
Big Bang theory (BBT), 29, 32, 41, 42, 46, 49, 135
Black holes, 44, 49, 82, 85, 88, 93–101, 103, 107–109, 111, 201, 202, 209, 235, 250
BL Lacertae objects, 42–44
Branes. *See* M-theory
Brownian motion, 65, 162–163, 234
Bulk. *See* M-theory

**C**
Casimir effect, 63, 99, 136, 237, 240
Category error, 155
C-field, 29, 40, 41
Cold spot (in CMB), 55, 122–124, 202
Complementarity, 138, 186, 187, 192
Complex numbers, 249, 251
Complex theory of relativity, 249–252
Copenhagen interpretation. *See* Quantum theory
Copernican principle, 27

Cosmic background explorer (COBE), 52, 55, 115, 118
Cosmic microwave background (CMB), 45, 46, 50, 52, 53, 55, 104, 109, 115, 117, 118, 120–124, 190, 202, 209, 210, 230, 244
Cosmic strings, 82–98, 103–105
Cosmological constant, 12, 65, 66, 258
Cosmological principle, 22, 27, 29, 119
Crab Nebula, 91, 92, 232
Curvature of space, 47, 78, 117, 235
Cyclic universe model, 19, 21, 135

**D**
Dark energy, 55, 213
Dark matter, 53–55, 176, 213, 245, 247, 253–255
Delayed action at a distance, 237
Design (evidence thereof in nature), 153
De Sitter horizon, 101, 102, 108, 109
De Sitter radiation, 101, 102
Doppler effect, 9, 12, 15, 23, 38, 243
Double quasar (Q0957+561A, B), 105, 106

**E**
Ekpyrotic theory of universe (Big Splat), 208–211
Electrons, 16, 17, 25, 63–65, 91, 98, 99, 136, 142, 163, 165–167, 170, 172, 173, 176, 193, 207, 240, 242, 243, 250, 251, 256, 257
Elements, formation of, 30, 32
Emission spectra, bright lines therein, 159, 166, 167
Entropy, 106–108, 251
Eons, 14, 23, 249–252
EPR paradox, 194, 195
Euclidean geometry, 47
Event horizon, 94–97, 99–101, 107, 108
Expansion of universe, 13, 15, 17, 19–21, 26, 29, 47, 49, 215, 241, 244

# Subject Index

**F**
Fine-tuning of universe, 144, 148, 152, 153, 203
Flattening of galactic rotation curves, 254, 255

**G**
Galaxies
 Andromeda (M31), 9, 11, 21, 28, 39, 75, 76, 126, 152, 217
 chains, 50, 52, 209, 224, 227
 clusters of, 13, 40, 103, 193
 compact groups, 219
 discrepant redshifts of, 214, 219, 220, 222–224, 227
 local group, 9, 20, 26, 126
 M33, 9, 11
 M87, 35, 37, 217
 magellanic clouds, 9, 252
 milky way, 8–10, 12, 20, 21, 27, 39, 52, 95, 127, 217, 225
 NGC 4319, 215–217
 and quasars, 33, 39, 41, 105
General relativity, 12, 29, 41, 56, 77–80, 84, 85, 88, 92, 110, 121, 209, 210, 234–236
Geodesic, 79
Gestalt psychology, 186, 189
God, 3, 8, 14, 66, 133, 150–155, 184
Gravitational lenses, 82–98, 103, 104, 121, 210, 227, 235
Gravitational waves, 52, 105, 109, 110, 209, 210, 230, 235
Graviton, 204
Gravity, 20, 21, 41, 55, 56, 61, 74, 77–83, 85, 86, 88, 94, 96, 99, 110, 127, 142, 201, 203, 204, 206, 207, 209, 229, 230, 233–237, 245, 246, 248, 254–257, 259

**H**
Hawking radiation, 99–101, 107, 136, 202
Heisenberg's uncertainty principle, 51, 63, 98, 102, 108, 137, 140, 194, 195, 237
Higgs particle, 66, 67
Hilbert's hotel paradox, 131, 132
Hologram, the universe as, 105–112
Hubble constant, 12
Hyperspace, 118, 173, 175, 192, 204, 206

**I**
Imaginary numbers, 249, 251
Imaginary unit, 249

Inertia, 59, 62, 64, 65, 67, 68, 140, 234, 236
Infinity, difficulty with concept thereof, 118
Inflationary theory, 52, 53, 122, 144, 210
Inflaton, 51

**K**
Kantian philosophy, influence on Bohr, 182
Kuiper's star, 89

**L**
Lamb shift, 63, 99, 136, 237, 240
Large Hadron Collider, 51, 66
Large number coincidences, 258–260
Laws of nature, 137–141
Le Chatelier universe, 21, 22
Light
 particle nature of, 22, 65
 velocity of, 23, 24, 28, 30, 46, 70, 71, 75, 95, 114, 145
 wave nature of, 163, 172
Lobachevskian geometry, 113
Lyman alpha forest, 44

**M**
Mach's principle, 21, 158
Magnetic monopoles, 50
Main sequence, 261
Mandl arcs, 84, 87, 220
Mass
 gravitational, 59, 60
 inertial, 21, 59–62, 64, 66, 67
Matrix algebra, 169, 170
Milesian school (of Greek philosophy), 1, 4
Modified Newtonian Dynamics (MoND), 255
Mother and baby paradox, 81
M-theory, 205–208, 211, 213, 229, 230
Multiple universes, 144–148, 202. *See also* Multiverse
Multiverse, 190, 193

**N**
Necessary and contingent being, 153–155
Neutrinos, 252, 253
 mega, 253
No hair conjecture, 94, 98
Nothing, ambiguity of concept, 137

Subject Index  271

**O**
Olbers' paradox, 125, 245
O-sphere, 125–130, 145, 241

**P**
Perfect cosmological principle (PCP), 29, 33, 42, 119
Picard topology, 116–118
Planck era, 258
Planck oscillator, 257
Planck satellite, 118, 119
Planck units
  area, 108, 110
  length, 108–110, 230, 232, 235, 256–258
  mass, 257
  temperature, 258
  time, 257, 258, 260
  volume, 257, 258
Planetary nebula, 89, 90
Plasma theory of universe, 242–245, 248
Polarization of CMB, 109, 210, 230
Positrons, 98, 136, 170, 193, 243
Principle of equivalence, 77, 78
Probability wave, 174–177, 189, 192
Ptolemaic model, 6

**Q**
Quantized redshifts, 227–229, 247
Quantum entanglement, 201–203, 251
Quantum theory
  Copenhagen interpretation, 180, 181
  many-worlds interpretation, 129, 146, 190
  subjective interpretation, 185
Quasars, 33, 36, 38–44, 72, 75, 76, 83–85, 97, 105, 106, 214–220, 246. *See also* Quasi stellar objects (QSOs)
Quasi stellar objects (QSOs), 44

**R**
Red giants, 33
Riemannian geometry, 47, 48, 113, 114, 139

**S**
Scale (as dimension of space), 13–18, 27–29, 47, 50, 54, 55, 59, 65, 77, 83, 89, 94, 119, 121, 138, 145, 193, 213, 227, 231, 244, 245, 247, 248, 253, 258

Schrodinger's cat (thought experiment), 177, 178, 180, 194
Second law of thermodynamics, 107
  generalized, 107
Seyfert's Sextet, 223–225
Shapley supercluster of galaxies, 121
Sloan Digital Sky Survey, 123
Special relativity, 22, 49, 71–74, 76–78, 161, 163
Spooky action-at-a-distance, 194–203
Standard model of particle physics, 66, 204
Steady state theory (SST), 21, 26, 27, 29–33, 38, 40–42, 44, 124, 214
  radical departure from, 40–44
Stephan's quintet, 219–222, 224
Stochastic electrodynamics, 236
String theory, 103, 123, 202, 204–207, 213
Supernovae, 31
Surface of last scattering, 45, 46

**T**
Thermodynamics, 106, 107
Time
  atomic, 23, 24
  clock, 23, 24
  dilation, 20, 72–75, 77, 78, 80–82, 95, 96, 102, 103, 249
Tired light explanation of redshift, 247
Topology of space, 50, 103, 117, 123, 213
Triple Alpha process, 149
Twin paradox. *See* Time, dilation

**U**
Universals, 1, 8, 13, 15, 20, 29, 47, 115, 124, 127, 128, 157, 241, 247

**V**
Vacuum
  false, 51, 52
  fluctuations, 62, 64–66, 99, 140, 234, 236, 240
Virtual particles, 63, 65, 67, 98, 99, 136, 237
VV 172 (galaxy chain), 224–227

**W**
Water (as primary substance), 4
Wave function, 173–176, 180, 189, 191–193
  physical interpretation thereof, 99

Weakly interactive massive particles (WIMPS), 54
White dwarfs, 54, 89–91
Wigner's friend (thought experiment), 178–180
Wilkinson Microwave Anisotropy Probe (WMAP), 52, 55, 115, 116, 118, 122
Wormholes, 201, 202

**Y**
Ylem, 16, 17

**Z**
Zitterbewegung, 65, 140, 234